THE ART AND SCIENCE OF BREAKTHROUGH RESEARCH

Unlocking Human Potential in the Age of AI

Jeremy Nixon

The Art and Science of Breakthrough Research: Unlocking Human Potential in the Age of AI

Jeremy Nixon

Table of Contents

Chapter 1

Mental Models of Genius: Understanding the Minds of Tesla, Einstein, and Other Visionaries

Great innovators do not simply float in an ocean of ideas, waiting for the perfect one to rise to the surface. Rather, they dive deep into the sea, masterfully navigating through the dark waters to sift through experiences, observations, past failures, and half-formed thoughts. Their uncommon intellect and tenacity have enabled them to make invaluable contributions to our world. Throughout history, visionaries in their respective fields have gifted us with groundbreaking ideas, technologies, and inventions that have shaped the world as we know it. Individuals like Nikola Tesla and Albert Einstein have left a profound impact on several disciplines, and the exploration of their minds can serve as an invaluable source of wisdom from which we can glean insights into our own problem-solving practices.

Tesla and Einstein not only stood at the pinnacle of technical mastery but were also known for their ability to synthesize complex information and use it to form a coherent understanding of the world around them. They employed mental models that combined knowledge from multiple sources, which helped them gain an in-depth conceptual grasp of the subject matter. Mental models are the representations of how an individual perceives, processes, and interacts with reality. For visionary thinkers, mental models went beyond simple compartmentalization of knowledge. They would encompass a wider variety of concepts, abstract ideas, and disciplines that were interwoven to form a rich tapestry of understanding.

Consider, for instance, Tesla's conceptualization of alternating current (AC) electricity. His background in electrical and mechanical engineering enabled him to recognize the potential of AC and develop a complete system for its generation, transmission, and usage. But it was his deep understanding of electromagnetism, a phenomenon that transcends the boundaries of traditional engineering, which allowed him to devise novel mechanisms and establish his revolutionary ideas. Thus, Tesla's mental model was a holistic blend of electrical, mechanical, and electromagnetism principles that enabled him to envision innovative solutions.

Similarly, Einstein's Theory of General Relativity, a profound accomplishment in the field of physics, showcases how his mental model was capable of integrating and augmenting ideas from diverse areas. Famously, his thought experiment of a person in a falling elevator suggested that gravity was indistinguishable from acceleration, leading to a radical reinterpretation of conventional concepts about mass, energy, and spacetime. Einstein's mental model broke away from the Newtonian paradigm to establish a new framework that enveloped the complex interconnectivity of matter, energy, and geometry. Both Tesla and Einstein show how mental models that are characterized by deep connections between different concepts can be crucial in crafting groundbreaking theories and innovations.

One common strategy for constructing such mental models involves using analogy and metaphor. Nikola Tesla often employed visualizations of his inventions before they were brought to life. This allowed him to test the efficacy of his devices and concepts in his mind, toying with the parameters and predicting outcomes in a purely theoretical plane before ever constructing them. Einstein also depended heavily on the visualization of physical phenomena, using analogy as a means of bridging the gaps between what was observable and what was deducible. The resulting mental models they constructed were able to capture the essence of underlying relationships between multiple concepts, leading to novel insights.

Developing and refining these mental models necessitated an ongoing commitment to expanding the knowledge base and updating one's beliefs regularly. Visionaries like Einstein and Tesla were voracious readers and relentless questioners, always probing the depths of their understanding to discover what might lie beneath. This process of questioning assumptions and seven unfounded beliefs with empirical evidence and logic allowed them to form new connections and generate innovative hypotheses.

In conclusion, the mental models of visionaries like Tesla and Einstein embody a deep understanding of the interconnected nature of reality and the vast implications of their insights. As we continue to dissect and learn from these cognitive roadmaps, we may uncover new avenues of innovation and hypothesis generation. We must strive to emulate these legends, not only in their insatiable appetite for knowledge but also in their capacity to integrate and synthesize ideas in novel ways. In doing so, we honor their legacies by making meaningful contributions to the advancement of human knowledge and ingenuity. As we venture into uncharted domains of research, we carry the torch of their wisdom, illuminating the path ahead and exploring the ever-expanding cosmos of intellectual discovery.

Introducing the Minds of Visionaries: Tesla, Einstein, and Beyond

The air was saturated with wonder, each particle infused with intense curiosity as two inquisitive, ambitious minds tackled the complexity of the universe in the late 20th century. These were no ordinary individuals. They were two of the most revered and innovative thinkers in history, forever linked by their groundbreaking ideas: Nikola Tesla and Albert Einstein.

Nikola Tesla - a prolific inventor and a creative polymath - was known for devising the alternating current (AC) electrical system, which powers the world we know today. But his innovations went far beyond this revolutionary achievement. Tesla pursued a wide array of subjects, from wireless communication to renewable energy. His intense imagination drove him to constantly test the limits of the world around him. Tesla's visionary mindset was not confined within the boundaries of failure or the oppressing opinions of critics - he dared to dream big, even when faced with ridicule.

In June 1900, Tesla asserted his belief that humans could effectively harness the force of gravity. This monumental claim was published in The Manufacturer and Builder magazine, where he outlined his idea for a machine that could control gravity and produce partial antigravity, an idea which even today seems fantastical. Tesla was heavily influenced by the concept of aether, believing that this mysterious substance could control all phenomena, from light to electricity and gravity. Although this belief led to intense criticism and disapproval, Tesla's ability to imagine and conceptualize such radical ideas only prove his exceptional visionary prowess.

As for Einstein, his prestigious theories of relativity, time dilation, and the photoelectric effect rewrote the rules of physics by unveiling a new understanding of the universe that transcended the prevailing norms of his time. His work on the famous equation E=mc^2 established a critical link between mass and energy, contributing significantly to the development of nuclear power. However, what truly set Einstein apart was his extraordinary capacity to formulate new scientific paradigms by considering the universe in an unconventional and creative manner.

An enlightening example of his visionary thinking can be found in his thought experiments, such as the "chasing a beam of light" experiment. This mental foray led to the development of the special theory of relativity. In one of his autobiographies, Einstein wrote, "A new idea comes suddenly, and in a rather intuitive way. But intuition is nothing but the outcome of earlier intellectual experience." This view highlights the role of his mental models, tools that frequently leveraged thought experiments to derive novel insights into the mysteries of reality.

Moreover, both Tesla and Einstein were masters in the art of associative thinking. Through seemingly unrelated connections, they discovered unexpected ideas and outcomes. Tesla, for example, would famously rely on mentally visualizing his inventions before physically building them, creating a complete mental prototype in the process. Einstein, on the other hand, would abandon conventional wisdom, blending the realms of space and time into a new, interdependent framework, which fundamentally altered our understanding of the cosmos.

While discussing renowned visionaries and intellects in history, it is essential to acknowledge the many figures who transcend the narrow confines of specialization, enriching our intellectual heritage with their polymathic knowledge and diverse accomplishments. From Leonardo da Vinci to Ada Lovelace, these polymaths captured the essence of a universe rife with interconnectedness, reshaping our perception of reality to better understand the intricate links between different disciplines.

In this vast mosaic of human brilliance, a key pattern emerges: the brilliance of great minds often arises not from their aptitude to conform to existing ideas, but to challenge them - relentlessly pursuing novel, revolutionary vistas of imagination. Through perseverance, resilience, and a steadfast determination to understand the enigmas of the natural world, visionaries such as Tesla and Einstein left indelible marks on our collective consciousness.

As our journey through the annals of human discovery unravels, let us consider this: what if we mastered the mental models and techniques deployed by these prodigious minds? What if we unearthed the secrets that produced world - changing ideas, adapting the cognitive architectures of visionaries to fit our contemporary context? Indeed, the audacious spirit of Tesla and the unbridled curiosity of Einstein echo through every word of this book, pushing us to embrace and learn from the mental techniques that propelled their genius, fervently challenging us to chart our own paths through the labyrinths of intellectual complexity, and driving us to seek the next groundbreaking innovation - all while unleashing the visionary within.

Unraveling the Uncommon Thought Processes of Genius Minds

There lies within the annals of human history a select pantheon of extraordinary minds, those who ventured beyond the boundaries of known science, philosophy, art, and human endeavor. They produced creations, concepts, and theories that forever transformed the trajectory of humanity and the very fabric of our understanding. How does one approach the study of these remarkable, genius minds? What cognitive processes and mental mechanisms allowed these visionaries to traverse such extraordinary intellectual terrain, where others dared not tread?

To embark on this quest for knowledge, we must delve into the complex architecture of their mental faculties, exploring their ability to perceive reality through a unique lens of creativity and innovation. Let us journey through a series of examples that elucidate the enigmatic thought processes of the genius minds, taking careful note of the intricate, interconnected workings of their creative power.

Consider the prodigious talents of brilliant composer Wolfgang Amadeus

Mozart. His ability to effortlessly conceive and compose music, as if by divine inspiration, left him with a unique mental capacity for grappling with intricate melodies and harmonies. Rather than merely exercising creativity and technical prowess, Mozart's mind was an immaculate machine. It was unconsciously capable of digesting vast amounts of auditory and musical information, re - arranging and synthesizing it in novel ways to produce some of the most enchanting symphonies and sonatas ever created. His ceaseless experimentation within the musical dimension endowed him with a unique ability to 'think in music,' allowing him to bend the boundaries of conventional composition in ways previously unimagined.

Similarly, consider the awe - inspiring talent of Leonardo da Vinci, whose artistic and intellectual accomplishments exemplify the epitome of the Renaissance man. Da Vinci's insatiable curiosity, his quest for interdisciplinary knowledge, allowed him to cultivate a diverse array of mental models. His mind would synthesize and integrate these diverse elements, creating unique, groundbreaking art and unprecedented inventions. For instance, da Vinci's study of human anatomy underpins his unparalleled depiction of the human form in paintings such as 'The Last Supper' or 'The Vitruvian Man.' Our present - day advancements in fields like engineering, optics, and air travel owe their gratitude to da Vinci's relentless exploration and vivid imagination.

Now, we turn our gaze to the towering intellect of theoretical physicist Albert Einstein who, through his thought experiments and mathematical prowess, permanently altered our understanding of space and time. His keen intuition, sharp focus, and ability to distill scientific truths through metaphorical and reductive reasoning led him to develop the iconic theory of relativity. Einstein's mental agility was a testament to his disregard and even rebellion against existing constructs and dogma. This enabled him to question the very fundamentals of the physical world and, consequently, create a conceptual framework that presaged the dawning of a new era in physics.

As we begin to ascertain the threads that bind these genius minds together, we must acknowledge their ability to seamlessly traverse multiple dimensions of thought and perception. They are masters of associative thinking, adept at identifying previously unseen connections between seemingly unrelated ideas or disciplines. This capacity for associative thinking not only underpins their exceptional creativity but also leverages an inherent cognitive flexibility that characterizes such genius minds. This sense of fluidity, of transcending traditional boundaries, is the crucible in which their remarkable ideas are forged.

Furthermore, we cannot disregard the relentless, disciplined nature of these individuals' pursuit of knowledge and discovery. They embody the spirit of perseverance, fueled by a ravenous curiosity and an unyielding belief in their potential to reshape our understanding of the world. This unwavering resolve, combined with their singular mental abilities, paved the way for their everlasting impact on the body of human knowledge.

As this exploration draws to its conclusion, we cannot help but stand in awe at the intricate workings of these visionary thinkers. Their unmatched capacity for intellectual insight, coupled with an intrepid spirit of inquiry, ensured their place as harbingers of epoch-making discoveries and creations. In a fitting tribute to their legacy, we turn our attention to the various mental models and techniques that have propelled their genius, as we strive to foster our intellectual prowess amid the boundless landscape of modern research.

Identifying the Mental Models and Techniques Used by Visionaries

An essential aspect of visionary thinking is the ability to think beyond traditional boundaries, to find connections between seemingly unrelated domains of knowledge, and to apply these connections to novel problem-solving scenarios. This lateral thinking approach allows for the generation of innovative ideas. For instance, Nikola Tesla brilliantly envisioned the concept of alternating current, while Einstein revolutionized physics by developing the theory of relativity. In both cases, these thought leaders transcended prevailing beliefs and scientific frameworks, breaking new ground in their respective fields.

It is also essential to recognize the role of imagination and intuition in the thought processes of visionaries. Tesla often claimed that he would visualize inventions in his mind, constructing and testing his designs in his imagination before building the physical prototypes. This ability to imagine potential scenarios and solutions can be a powerful tool for effectively tackling complex problems. Similarly, Einstein has been famously quoted as saying that "imagination is more important than knowledge." He believed that having a curious, imaginative mind enabled him to explore new frontiers in science and form breakthrough theories like the General Theory of Relativity.

Another crucial technique employed by visionaries is the capacity for abstraction and pattern recognition. This process involves distilling complex phenomena and data into simpler and more fundamental elements, identifying patterns and trends, and extrapolating the bigger picture. By abstracting information and focusing on these basic elements, visionaries can reveal deep insights into the workings of a complex system, such as understanding the fundamental structure of the universe or the behavior of particles at subatomic levels.

Critical thinking and logical reasoning are also indispensable elements in the repertoire of great thinkers. By dissecting problems, identifying and challenging assumptions, and exploring alternative possibilities, visionaries like Galileo, Newton, and Darwin have developed groundbreaking theories that have reshaped our understanding of the world around us. Similarly, contemporary thinkers like Elon Musk and Richard Branson apply critical thinking and logical reasoning methods to envision and create innovative companies, products, and solutions.

Combining multiple disciplines is another powerful approach employed by visionaries throughout history. From polymaths like Leonardo Da Vinci to creative modern pioneers like Steve Jobs, the ability to integrate insights and ideas from various domains of knowledge has led to revolutionary discoveries and creations. By integrating seemingly unrelated disciplines, these innovative minds have paved the way for breakthrough technologies and inventions that have transformed the course of human history.

Lastly, the capacity for persistence and resilience in the face of failure and adversity is a fundamental characteristic of visionary thinkers. As Thomas Edison famously said, "I have not failed. I've just found 10,000 ways that won't work." This indomitable resolve to push through setbacks and continue striving for success has resulted in momentous discoveries and breakthroughs.

By examining the mental models and techniques of visionaries, it becomes evident that fostering lateral thinking, imagination, intuition, abstraction, critical thinking, logical reasoning, interdisciplinary knowledge, and resilience are crucial components to nurturing a creative and revolutionary mind. Aspiring to emulate these characteristics and techniques, we can harness the power of visionary thinking and contribute to the constant evolution of human knowledge and progress. Our journey, much like that of the visionaries before us, is not a solitary pursuit but a collective endeavor to push the boundaries of knowledge, challenge the status quo, and develop a deeper understanding of the world and our place within it. In this journey, each new insight lights the path forward, paving the way for new explorers, intent on discovering the next frontier.

Adapting and Implementing Visionary Mental Models into Modern Research

Einstein's ability to merge visualization and mathematics allowed him to create groundbreaking theories in physics. He strongly emphasized the importance of visual thinking in grasping and simplifying complex concepts. Anecdotes describe how he would imagine himself riding a beam of light to understand the relationship between space and time, which ultimately helped him deduce the theory of relativity. Borrowing from Einstein's visionary mental model, today's researchers can hone their visual thinking skills through deliberate practice and employment of rapid prototyping techniques during problem-solving.

Similarly, Nikola Tesla was undeniably ahead of his time and credited with conceptualizing and designing modern alternating current (AC) electricity systems, among several other inventions. He believed in the power of visualization, imagination, and frequent daydreaming to connect abstract ideas. Tesla's mental models permitted him to run entire experiments in his mind, predicting outcomes with impressive accuracy before conducting them in reality. This ability saved time, energy, and resources during the research process. Modern researchers can benefit from systematically training their minds to engage in more imaginative, visionary thinking and develop mental representations aligned with their research objectives.

Moreover, both Einstein and Tesla's pursuit of knowledge extended beyond the boundaries of their respective domains, straying into interdisciplinary territory. Historical evidence suggests that Einstein's violin playing and Tesla's interest in poetry, literature, and global politics stimulated their creative and abstract thinking processes. Incorporating a cross-disciplinary approach allows researchers to formulate unconventional solutions and identify unique patterns across different fields. Acknowledging the importance of polymathy in contemporary research, interdisciplinary collaborations that foster innovative ideas can lead to a collective acceleration of knowledge generation.

A testament to the rigorous open-mindedness of great thinkers is their approach to assumptions and biases. Visionaries like Einstein and Tesla were willing to question existing beliefs and cross-examine their own theories to identify logical fallacies. The inclination to pierce through biases and adopt first principles reasoning encourages researchers to examine the root of a problem and craft innovative solutions from scratch, often leading to disruptive breakthroughs. Creating a conducive atmosphere for debates, discussions, and constructive criticism within research teams can encourage open-mindedness and absence of dogma in solving complex problems.

When considering adapting and implementing visionary mental models, it is crucial to understand that every individual's cognitive style and personal experiences shape their thinking processes. Some of the essential aspects of adopting mental models include nurturing creativity, developing strong visualization skills, embracing interdisciplinary thinking, dissecting assumptions, and fostering objectivity. These aspects can be infused organically into a researcher's methodology to suit one's unique cognitive abilities and preferences.

In conclusion, by tailoring and assimilating visionary mental models into their research routines, modern researchers unlock the door to transcendent ideas that may shape our world. Embracing a sense of childlike curiosity and wonder can allow researchers to surpass limitations imposed by conventional thinking and transform seemingly insurmountable obstacles into stepping stones. The road to discovery is neither straight nor without setbacks, but the enduring footprint of stalwarts like Albert Einstein and Nikola Tesla serves as a beacon of inspiration and motivation; pushing boundaries and inspiring generations to dream beyond imagination, reaching for the stars.

Chapter 2

Hypothesis Generation: Leveraging Associative Networks for Groundbreaking Ideas

The landscape of scientific breakthroughs is vast and mysterious, guided by a process of questioning, observing, and testing our understanding of the world around us. At the heart of this process lies a crucial step: hypothesis generation. It is in this remarkable cognitive realm that groundbreaking ideas are born, driving innovation and shaping our perception of reality.

To gain an appreciation for the significance and power of associative networks in hypothesis generation, we must first venture back in time to visit the inquisitive mind of Charles Darwin. As he studied the diverse species inhabiting the Galápagos Islands, Darwin built upon the works of other naturalists and geologists, constructing a vast associative network of knowledge. Through these connections, he was able to develop the then-revolutionary idea of natural selection. Darwin's associative network granted him the capacity to view the natural world through a fresh lens, fundamentally reshaping our understanding of evolution.

Central to the concept of associative networks is the idea that the human brain is an intricate web of interconnected ideas, memories, and pieces of information. Each experience, thought, and observation that we encounter becomes a node in this network, with stronger connections forming between nodes that are frequently activated or related. As we weave together various strands of our associative network, we become better equipped to generate novel ideas, formulating fresh hypotheses that challenge the status quo.

To bolster the power of associative networks in driving hypothesis generation, it is essential first to nourish our cognitive landscape with diverse experiences and knowledge. As we expose ourselves to a broad range of disciplines, perspectives, and ways of thinking, we equip our minds with the necessary ingredients to assemble groundbreaking hypotheses. To effectively leverage associative networks, we must also learn to build connections between seemingly unrelated areas, recognizing patterns and parallels where others may not.

Consider the remarkable examples set by many historical geniuses and visionaries, such as Leonardo da Vinci and Nikola Tesla. These individuals possessed the unique ability to see connections between disciplines that would initially appear disparate, weaving together threads of knowledge from multiple fields to generate novel ideas. Inspired by the flight mechanisms of birds, da Vinci, in pursuit of human flight, theorized the possibility of creating a device that could translate human muscle power into aerodynamic lift - the result: his sketches of the ornithopter, an early conceptualization of a helicopter.

In the same spirit, Tesla's insights into electromagnetic fields resulted from his deep understanding of natural phenomena and physics, as well as his ability to visualize complex structures. By deftly navigating his extensive associative network, Tesla produced inventions that would forever change the course of global energy production, such as the alternating current induction motor and transformer.

There are, of course, challenges and constraints to developing and utilizing associative networks in hypothesis generation. As humans, we are often hindered by cognitive biases, dogmas, and a fear of the unknown. However, concrete steps and strategies can help overcome these obstacles in hypothesis generation. For instance, engaging with other individuals with diverse perspectives, backgrounds, and expertise can expose us to new information and modes of thinking, thereby enriching our associative networks.

Another potent approach is to engage in activities such as brainstorming or thought experiments, which can help spur lateral thinking and stimulate connections within our associative networks. For example, Albert Einstein's famous thought experiments on light and time led to his development of the Theory of Relativity, forever altering our understanding of space and time.

In conclusion, the true potential of human ingenuity awaits at the intersection of knowledge, experience, and innovation. By nourishing our minds with a diverse array of information and embracing the power of associative networks, we can drive groundbreaking ideas and redefine the boundaries of human understanding. This same curiosity which led Darwin to venture to the Galápagos Islands, da Vinci to contemplate human flight, and Tesla to envision alternate paths of energy production, lives on within the ambitious researchers and innovators of the world. The journey of discovery continues, as each new observation, question, and insight adds another thread to the intricate tapestry of investigation, bound together through associative networks, transcending disciplinary boundaries, and ultimately leading to a brighter, more informed future for us all.

Introduction to Associative Networks: The Science Behind Hypothesis Generation

As we delve into the exploration of genius minds in effervescent fields of scientific inquiry, it is useful to first understand the molecular structure and mechanics behind their unparalleled ability to generate innovative and groundbreaking hypotheses. Let us take a fascinating look at the intricate web of "associative networks" in the brain, the foundation upon which the thought processes and mental frameworks of inquisitive scholars and masterful intellectual creators are built.

Associative networks are complex organizations of interconnected ideas, memories, sensations, and experiences that form through the brain's robust capacity for pattern recognition and information processing. When a thought or an observation emerges in our consciousness, it is often connected to a myriad of other related thoughts and experiences through a series of neural pathways and associations. Essentially, our brains excel at connecting the dots, finding links between seemingly disparate pieces of information, and weaving them together to create new, innovative hypotheses and ideas.

Consider the example of Sir Isaac Newton when he famously saw an apple falling from a tree and had a moment of enlightened intuition that eventually led to the development of the Universal Law of Gravitation. It was not the mere observation of the fallen fruit that sparked Newton's intellectual breakthrough but rather his exceptional ability to connect that observation to his existing knowledge of physics and mathematical principles. His associative networks potentiated the birth of a radically new scientific understanding of the forces that govern the physical universe.

In recent years, advancements in neuroscience and cognitive psychology have revealed crucial insights into how associative networks are formed and strengthened in the brain. Researchers have shown that the formation and activation of associative networks are intimately connected to the brain's capacity for "neuroplasticity," or its ability to alter the structure of its own neural connections in response to new experiences and evolving environmental demands. The more robust and diversified one's associative networks are, the greater their ability to generate innovative and transformative hypotheses.

When the minds of visionaries such as Nikola Tesla or Albert Einstein were confronted with seemingly insurmountable scientific conundrums, their incredibly potent associative networks allowed them to synthesize information in novel ways, simultaneously drawing upon diverse fields of expertise and integrating seemingly unrelated pieces of data to construct powerful new lines of reasoning. This ability to tap into vast reserves of interwoven ideas and layer them in original configurations was a driving force behind their paradigm - shifting contributions to human knowledge and achievement.

But how might we, as aspiring intellectual explorers ourselves, emulate these luminary thinkers to strengthen and expand our own associative networks? One technique to consider is the deliberate cultivation of multidisciplinary knowledge and curiosity, immersing oneself in a broad range of fields and experiences. This helps build a heterogeneous repertoire of ideas and perspectives that can be subsequently integrated and recombined in the service of generating innovative hypotheses. As the frontier of human knowledge advances further, this expansion of associative networks will be increasingly crucial to developing transformative breakthroughs and insights.

The power of the associative networks, however, should not be confused with the archaic idea of a "eureka" moment or an epiphany of divine inspiration. These intellectual leaps are not sudden or random gifts from the universe; rather, they are an emergent property of the robust, intricate, and interwoven tapestry of ideas, memories, and experiences that form the backbone of our cognitive universe. As we refine our associative networks and plunge into the fertile depths of human ingenuity, we become primed to unfold the potential that lies hidden in the folds of our mental landscape and harness the secrets that await our discovery.

These very principles can be captivatingly observed in a hypothetical scenario where Tesla and Einstein confer on a scientific problem. The interplay of associative networks between these two brilliant minds might have resulted in an explosion of intellectual synergy, generating insights and ideas that have pushed the boundaries of human understanding even further. This vivid illustration of the power of associative networks, alongside the many historical examples that dot our scientific annals, serves as a tantalizing reminder of the untapped potential that lies dormant within each one of us, awaiting the spark of curiosity to ignite our mind's boundless creativity.

Building Associative Networks: Techniques and Strategies for Strengthening Connections

Building associative networks is a powerful means of strengthening connections between seemingly unrelated concepts, ultimately yielding new insights and driving innovation. Associative thinking is a key mechanism behind hypothesis generation, as it involves identifying patterns and creating new connections between ideas drawn from diverse sources. By implementing various techniques and strategies to enhance associative networks, researchers can augment their creativity, overcome cognitive biases, and develop groundbreaking ideas.

To build an associative network, the first step is enhancing knowledge absorption. Developing a habit of continuous learning across various disciplines is crucial in creating an environment for making new associations. One can achieve this through reading books, attending lectures and conferences, engaging in discussions, or taking online courses in diverse subjects. By doing so, a researcher acquires a profound understanding of different domains, which occasionally intertwine to produce innovative ideas. For example, computer science pioneer Alan Turing blended his extensive knowledge of mathematics, logic, and biological processes to develop the foundation of modern-day computer systems.

In addition to acquiring knowledge, it is essential to routinely engage in reflection and mental organization. Allocating time each day for mindful activities like meditation or journaling allows the mind to absorb and process the vast array of information from books, lectures, and conversations. It is during these moments of quiet reflection that associations between seemingly unrelated ideas emerge, often leading to groundbreaking insights.

Another technique that significantly enhances associative thinking is mind mapping. This visual tool enables researchers to organize, prioritize, and analyze ideas by graphically representing the relationships between concepts. For instance, mind maps could depict the connections between various scientific theories or historical events, allowing the researcher to identify common threads and patterns among seemingly divergent topics. This visual representation of related ideas can lead to deeper understanding and rich connections, further enhancing the process of hypothesis generation.

The deliberate and systematic practice of brainstorming is another crucial aspect of building associative networks. Arranging dedicated sessions for generating hypotheses encourages the mind to relentlessly explore new ideas and connections. The key to successful brainstorming is to avoid self-censorship and judgment; instead, researchers should aim to cultivate an open-minded environment where all ideas flow freely. To produce a diverse range of hypothesis candidates, one could employ various brainstorming techniques, such as analogical thinking, which involves examining a problem through the lens of an unrelated domain, or lateral thinking exercises that challenge conventional assumptions.

When undertaking associative thinking exercises, it is vital to maintain a balance between exploration and exploitation, ensuring that fresh connections are actively sought without getting lost in the maze of possibilities. To achieve this balance, researchers can alternate between periods of focused work (exploitation) and relaxation or engaging in unrelated activities (exploration). For example, Charles Darwin reportedly followed a strict daily routine, which included dedicated time for walking and reflecting on his ideas. These moments of introspection and relaxation may have fostered mental agility, helping Darwin uncover groundbreaking connections and produce his revolutionary theory of evolution.

A crucial element in building associative networks is cultivating a mindset of curiosity and openness to new experiences. As researcher Isaac Asimov once remarked, "The most exciting phrase to hear in science... is not 'Eureka!' but 'That's funny...'". Embracing serendipity and maintaining a willingness to explore peculiar ideas is invaluable in nurturing associative thinking. A famous instance of serendipity yielding a breakthrough discovery is the development of the microwave oven; while working on a radar-related research project, American engineer Percy Spencer accidentally discovered that microwaves could heat food.

In conclusion, the art of building associative networks relies on several key techniques and strategies. By fostering continuous learning, engaging in reflection and mind mapping, cultivating curiosity, and maintaining a balance between exploration and exploitation, researchers can unlock the power of associative thinking and catalyze groundbreaking discoveries. From the inspirational visions of Tesla and Einstein to the computational breakthroughs of Turing and the serendipitous invention of the microwave, associative thinking has been the driving force behind many revolutionary ideas that have shaped human history. With curiosity and tenacity, researchers can harness this power and continue the legacy of groundbreaking research.

Case Studies: Historical Examples of Successful Hypothesis Generation through Associative Networks

One of the most brilliant minds of all time, Leonardo da Vinci, exhibited what we now recognize as the associative thinking process. His inquisitiveness often led him to examine the deeper mysteries of nature. Upon witnessing the majesty of birds taking flight, Leonardo's attention turned skyward, seeking to uncover the secrets of flight. He devoted countless hours to studying and sketching birds, analyzing their wings and the fluid movements of their bodies. This intuitive ability to analogously relate the function of mechanical devices to natural phenomena allowed him to conceive novel ideas for flying machines, which were, at the time, considered unthinkable and unattainable. While his designs were ultimately limited by the technological constraints of his era, the core principles behind his inventions foreshadowed the development of modern aviation.

Another admirable case of successful hypothesis generation through associative networks is the pioneering work of Gregor Mendel. As curiosity of the world led him to train his eye on the humble pea plant, he sought to discover the hidden mechanisms underlying patterns of inherited traits. Over many years, he meticulously crossed and hybridized hundreds of pea plants, carefully observing and tracking the traits revealed in each new generation. Through his systematic approach, Mendel discovered the fundamental laws of genetics. His work laid a foundation that modern scientists built upon, ultimately leading to our current understanding of the complex interplay of genes and the environment in shaping an organism.

Louis Pasteur's groundbreaking work on the germ theory of disease also serves as an excellent example of associative thinking. Observing that the wine and beer placed in corked vessels did not spoil, Pasteur hypothesized that microorganisms in the air were responsible for fermentation and the subsequent spoiling of food. By designing his now-famous swan-necked flasks, he demonstrated that when the flask's neck was intact, no contamination would occur. However, when the neck was broken, allowing air and microorganisms unrestricted access to the liquid within, they rapidly multiplied, spoiling the contents. Pasteur's elegant experiment unveiled the link between microorganisms and fermentation and ultimately led to the development of pasteurization, the life-saving process that has preserved countless food products and warded off diseases for well over a century.

The work of Michael Faraday provides yet another illustration of hypothesis generation through associative networks. Faraday was not a formally educated man; nevertheless, his boundless curiosity and innate ability to identify connections between diverse phenomena enabled him to lay the foundation for modern electromagnetism. In a series of remarkable experiments, Faraday demonstrated the interrelatedness of electric and magnetic fields. Specifically, he explored the reciprocity between magnetic fields and electric currents in what is now known as electromagnetic induction. His experiments, which elegantly illustrated the principles of creating magnetic fields around conducting materials, laid the groundwork for the modern electric power infrastructure.

These examples provide a testament to the immense power of associative thinking. Each of these individuals, though diverse in background, interest, and time, were united by their ability to identify connections between seemingly disparate ideas, leading them to propose groundbreaking hypotheses. By embracing this associative network-based mode of thinking, they expanded the reaches of scientific knowledge, reshaped the way we conceptualize our world, and prompted us to ask deeper questions about the nature of existence.

Overcoming Obstacles: Addressing Challenges and Limitations in Hypothesis Generation

Throughout human history, great thinkers have developed powerful ideas that reshaped the landscape of knowledge. However, none of these visionaries ever embarked on their journey without any obstacles. Hypothesis generation, as a crucial phase of scientific exploration, always comes with its fair share of challenges and limitations. Overcoming these hurdles is essential to cultivating the mind of a visionary and propelling oneself to the forefront of scientific discovery and innovation.

At the core of hypothesis generation lies the need to ask pertinent, well-crafted questions. A common challenge faced by researchers is formulating concise, focused inquiries that are capable of generating useful and fruitful avenues of investigation. This often stems from a lack of clarity in the problem at hand, resulting in vague and unproductive hypotheses that fail to inspire further investigation.

To address this challenge, researchers must actively engage in the process of defining the problem they hope to investigate. This involves breaking down complex ideas into more manageable components and identifying key variables and relationships. By enunciating the boundaries and components of the problem, researchers can better articulate the questions they seek to answer and avoid the pitfall of unfocused hypothesis generation.

Another challenge in hypothesis generation is overcoming cognitive biases. These mental shortcuts often lead individuals to favor certain lines of reasoning or conclusions, potentially leading to incomplete or erroneous hypotheses. In order to sidestep these cognitive biases, researchers can practice techniques such as the "devil's advocate" approach, wherein they actively challenge their own ideas and assumptions. By welcoming scrutiny and critique from others, researchers can refine their hypotheses and incorporate diverse perspectives.

Distractions and information overload can also hinder the creative process of hypothesis generation. With an ever-growing body of knowledge and a constant barrage of data, researchers can become easily overwhelmed by the sheer volume of information they encounter. However, by employing strategies such as pruning irrelevant information and focusing on core insights, researchers can navigate these cognitive obstacles successfully. One effective technique is the creation of conceptual maps, where ideas are visually represented, helping untangle complicated concepts and revealing hidden connections.

A further challenge in hypothesis generation lies in the weight of the existing paradigms and dogmas. Scientific researchers often operate within the confines of prevailing theories, models, and worldviews, which may limit their ability to entertain alternative hypotheses or challenge established beliefs. This is particularly problematic in interdisciplinary research, where assumptions and methods from one field may not hold in another.

To overcome this challenge, researchers should adopt a flexible, open-minded approach to problem-solving that actively embraces paradigm shifts and encourages the exploration of unconventional ideas. This can involve taking inspiration from other fields or seeking out knowledge beyond one's area of expertise, promoting the exchange of ideas and fostering innovation.

Finally, hypothesis generation can be stymied by a dearth of creativity and the presence of rigid thinking patterns. Human brains are wired to seek out patterns and develop mental models to simplify complex problems. While this efficiency can be valuable, it can also lead to pitfalls - especially when these mental models become outdated or overly simplistic.

To surpass these barriers, researchers should engage in activities that promote creativity and divergent thinking. This can take the form of brainstorming sessions, guided imagery exercises, or even seeking artistic inspiration from different disciplines. Furthermore, mindfulness practices, such as meditation, have been shown to improve cognitive flexibility, fostering a state of mind conducive to generating fresh insights and innovative ideas.

In conclusion, hypothesis generation is no stranger to obstacles and limitations. However, by confronting these challenges head-on and employing deliberate strategies to overcome them, researchers can unleash their inner visionary, generating novel hypotheses that catalyze groundbreaking discoveries. As we progress through the ever-evolving landscape of scientific research, the importance of cultivating these skills only becomes more vital - for it is through the intelligent, persevering navigation of the unknown that new constellations of human understanding are forged.

Practical Exercises and Quizzes: Training the Brain to Leverage Associative Networks

As researchers and innovators seek to access the untapped potential of associative networks, it is essential to engage in practical exercises and quizzes that train the brain to leverage these mental pathways. By harnessing the power of associative thinking, one can forge unforeseen connections between seemingly disparate ideas - a key driver of groundbreaking discovery.

One practical exercise for cultivating associative networks is the 'word ladder' game in which individuals generate associations between two given words. For instance, taking the words 'car' and 'elephant,' the objective would be to form a chain of words, each one associated with the previous word. An example might be: car, road, journey, safari, wildlife, elephant. As a variant of the exercise, one can substitute words with concepts or problems in specific research domains. The aim is to build mental agility and encourage the pursuit of associations that spawn innovative ideas.

Another useful exercise to develop associative thinking is 'random input generation,' which can incorporate concepts from several fields of knowledge. Researchers can create a list of prompts, each drawn from a different discipline. The aim is to find connections between the prompts that spark novel hypotheses. For instance, prompts such as 'impressionist painting,' 'wind energy,' and 'swarm intelligence' could lead to the concept of a wind farm with turbines designed based on the motion patterns of flocking birds to maximize efficiency. Challenging oneself to synthesize information from diverse fields can fortify associative networks and in turn, accelerate hypothesis generation.

A priming exercise for associative networks also serves as an excellent training tool. This technique involves presenting a series of images to an individual and asking them to generate hypotheses that incorporate all the images. For instance, a series including a lightbulb, a clock, and a river might lead to the idea of harnessing flowing water as a power source to control lighting based on a predetermined schedule, which could be of interest for off-grid lighting solutions. Such exercises expose one's brain to multiple ideas in succession, encouraging the formation of associations and strengthening cognitive pathways necessary for creative problem-solving.

Priming quizzes can be an effective means of training the brain to accelerate its hypothesis-generating muscle. These quizzes consist of a series of questions that prompt researchers to actively engage with new concepts and theories. For example, questions can range from identifying similarities between unlikely duos (e.g., Mozart and an enzyme) or envisioning how future technology might merge with nature (e.g., bio-inspired transportation systems). By prompting critical thinking and creative problem-solving skills, these quizzes sharpen the capacity for associative thought and catalyze innovative solutions.

Mental exercises involving skillset swapping can also prove beneficial for nurturing associative networks. Researchers should imagine they possess the skillset of a famous expert from an unrelated field and apply it to their own research domain. For instance, a biologist might adopt the perspective of a professional dancer and evaluate their research questions through this unconventional lens. Potentially, this could lead the biologist to consider the rhythmic aspects of biological systems, thus generating fresh insights impact cellular mechanisms. The fusion of these distinct perspectives fosters novel associations and ultimately, paves the way for groundbreaking discoveries.

Engaging in these exercises and quizzes regularly not only strengthens associative networks but also provides a fresh perspective on existing research problems. By fusing new ideas and forging interdisciplinary connections, researchers can unlock the doors to uncharted territories of knowledge.

In the tradition of pioneering visionaries such as Einstein and Tesla, those who adopt these associative thinking techniques will be well equipped to probe the depths of human understanding. The constant search for connections, combined with the steadfast pursuit of intellectual diversity and interdisciplinary cooperation, primes researchers for challenges that lie at the forefront of innovation. The lessons derived from these exercises thus fuel the onward journey into the unexplored realms of human achievement, forging a path towards revolutionary discoveries that will reshape the world as we know it.

The Role of Associative Networks as a Tool for Driving Groundbreaking Ideas and Future Research

In understanding the role of associative networks in driving groundbreaking ideas and future research, we need to appreciate the nuances and subtleties in the working of the human mind. Oftentimes, the most profound scientific discoveries arise not from solitary moments of brilliance, but from the active cultivation and deliberate leverage of associative networks.

Associative networks consist of interconnected concepts and ideas that form mental cognitive structures in our brains. These structures underpin and facilitate the subconscious formation of new hypotheses, connections, and insights that lead to groundbreaking discoveries. By embracing the heuristic power of associative networks, we can enhance our creativity and problem - solving capabilities, invigorating the research process and accelerating the pace of scientific advancement.

Consider the case of James Clerk Maxwell, the 19th - century physicist who made significant contributions to the fields of electricity and magnetism. Maxwell was known for his strong associative network-his vast mental library of mathematical concepts, scientific principles, and empirical observations. By accessing and leveraging these interconnected nodes of knowledge, he devised groundbreaking mathematical models of electromagnetic waves, setting the stage for the modern era of telecommunications and electronics. Maxwell's work exemplifies the power of associative networks as tools for driving innovation and creative problem - solving.

Another example, closer to our modern world, is the work of Kary Mullis, who won the Nobel Prize in Chemistry in 1993 for the invention of the polymerase chain reaction (PCR) technique. This method has become a cornerstone of molecular biology research, enabling scientists to amplify specific DNA segments exponentially. Mullis was inspired to develop PCR while driving through the Californian countryside, reflecting on the complexities of DNA replication. This scientific breakthrough was made possible by Mullis's robust associative network of biological knowledge - his understanding of the principles of molecular biology coupled with his grasp of chemistry and biochemistry.

These examples serve to highlight the critical importance of associative networks in the genesis of groundbreaking ideas and scientific discoveries. To harness the true potential of associative networks, we must adopt a holistic approach to learning and research. This approach entails fostering connections between seemingly unrelated disciplines while honing our mental faculties to recognize subtle links and patterns.

However, the path to innovation is fraught with challenges, and associative networks are no exception. Overreliance on these mental structures can potentially stifle creativity and hinder our ability to think outside the box. To mitigate these risks, we must learn to balance the use of associative networks with other cognitive tools.

For instance, brainstorming - a popular method for generating new ideas - encourages the free association of thoughts, synonymous with the workings of associative networks. However, many critical breakthroughs may also arise from periods of focused, deliberate thought - an approach known as 'convergent thinking.' By balancing both 'divergent' (brainstorming) and 'convergent' thinking, we strike a balance between free - flowing creative exploration and steadfast logical analysis.

Moreover, the role of external stimuli and sensory input in associative networks cannot be overstated. Associations are not formed in isolation but are products of the interplay between the mind and the surrounding environment. By seeking diverse sensory experiences and engaging with rich stimuli, we add building blocks to our associative networks and enhance their potential to yield novel insights.

The development of strong associative networks demands a lifetime of dedicated learning, curiosity, and exploration. However, harnessing their power need not be a Herculean task. A few simple techniques - such as expanding one's knowledge base, engaging the senses, and promoting a conducive environment for creative exchange - can prime the mind for unexpected discoveries and transformative ideas.

As we look to the future of research and innovation, we are reminded of the Latin phrase "exprime melius," which translates to "draw out the best." The challenges that lie ahead require us to catalyze our collective cognitive capacities and engage our associative networks - leveraging the power of the mind's inherent complexity. By doing so, we stand poised to unlock groundbreaking ideas and usher in a new era of scientific discovery. In the immortal words of Isaac Newton, "If I have seen further, it is by standing on the shoulders of giants." Let us embrace the power of associative networks and strive to look further than ever before, unlocking the potential that lies within the intricate fabric of human thought.

Chapter 3

Visual Thinking Techniques: Rapid Prototyping for Successful Research

One of the most famous instances of visual thinking comes from Tesla, who claimed he could picture the complex inner workings of his inventions in his mind's eye with vivid detail. This innate ability to manipulate mental images enabled Tesla to perform experiments in his imagination before actualizing them in reality. Although most of us may not possess such a natural gift, there are practical techniques through which researchers can hone their visual thinking skills.

One of these techniques is sketching. Sketching is often associated with artistic endeavors, but its role in science and research should not be underestimated. Sketching provides a means to externalize thoughts, freeing up cognitive resources for higher-level thinking and problem-solving. For example, Charles Darwin often sketched his ideas and observations while sailing on the HMS Beagle, which ultimately helped him formulate the theory of evolution. Likewise, Santiago Ramón y Cajal, the father of modern neuroscience, was inclined towards sketching intricate details of neurons he observed under the microscope. By giving their ideas a physical form, these researchers facilitated the refinement and organization of their thoughts, fostering a deeper understanding of their work.

A related technique is the creation of visual maps, which depict the relationships between different pieces of information in a spatial layout. These maps serve as a landscape for researchers to explore the connections among ideas, encouraging the discovery of novel insights. Consider the work of renowned physicist Richard Feynman, who utilized a graphical framework called "Feynman diagrams" to visualize the interactions of subatomic particles. The diagrams proved essential in conceptualizing complex phenomena and generating new hypotheses. Visual mapping techniques can easily be adapted to any field, from molecular biology to social sciences, to strengthen understanding and uncover unseen connections.

In addition to these 2D representations, researchers may benefit from rapid prototyping techniques that involve creating three-dimensional models or digital simulations of their work. Isambard Kingdom Brunel, a famed British engineer, frequently used scale models of his designs for bridges and tunnels to develop a better understanding of their practical considerations. Today, researchers can harness digital technologies like 3D printing and virtual reality to efficiently experiment with physical prototypes and refine their ideas.

An important aspect of visual thinking is the willingness to iterate and embrace the imperfect nature of a prototype. Fear of failure or perfectionism can inhibit the creative process, impeding the development of innovative ideas. However, fruitful visualization techniques thrive on the flexibility and openness to change. It is through iteration and refinement that a rough sketch or model can evolve into a profound, transformative insight.

In conclusion, researchers of any discipline can attain a wealth of benefits from incorporating visual thinking techniques into their work. By generating sketches, maps, physical models, or digital simulations, mental barriers are broken down, and fresh perspectives are illuminated. The art of rapid prototyping engenders a symbiotic relationship between the mind's capacity for analysis and its desire for creativity, cultivating an environment that fosters breakthrough achievements. As researchers look towards the future, envisioning the course of human progress, it becomes evident that visual thinking will remain an invaluable tool. By honing these techniques, we can empower ourselves to traverse the landscapes of possibility, unveiling the hidden connections and patterns that shape the world around us.

Introduction to Visual Thinking: How Rapid Prototyping Enhances Research

The concept of visual thinking in research can be thought of as the creation of mental images, diagrams, or simulations of a problem, which can then be used to test, analyze, and refine potential solutions. The tangible embodiment of this idea is rapid prototyping, which entails the construction of physical or digital representations of ideas for the purpose of testing and iterating on them quickly.

Visual thinking is extraordinarily valuable in the research process because it allows us to externalize our mental models and make abstract concepts more concrete. This process enables us to engage with complex ideas and see how they might interact with real-world systems, giving us a clearer sense of what will work and what won't. Moreover, by examining and analyzing our visual prototypes, we can generate new ideas, identify gaps in our understanding, or question our assumptions, leading to an overall more robust and innovative solution.

Consider, for instance, Albert Einstein's development of the theory of relativity. During his thought experiments, Einstein would visualize himself riding alongside a beam of light, witnessing hypothetical scenarios that defied conventional interpretations of time and space. This technique enabled him to work through complex conceptual puzzles, ultimately concluding that the speed of light is constant, regardless of the observer's motion. Without a doubt, these visualizations played an integral role in the refinement of his ideas and provided a solid foundation for his groundbreaking research.

Another notable example comes from Nikola Tesla, who was known to have a unique visual thinking process. Tesla could mentally generate a vivid 3D image of his inventions and manipulate them to test their functionality. He referred to this technique as his "mind's workshop," where he would rapidly prototype and iterate on his designs without ever requiring a physical model. This ability to visualize and analyze his ideas enabled Tesla to perfect his work on alternating current motors, high-frequency electrical oscillators, and many other inventions.

To harness the full potential of visual thinking in our own research, it's essential to adopt specific techniques and strategies for sketching, mapping, and prototyping ideas. Taking inspiration from historical visionaries, we can experiment with various methods such as mind maps, system flowcharts, or rough sketches to externalize abstract concepts and uncover hidden connections within our research. Importantly, by embracing simplicity and speed, we make it easier to iterate and refine our prototypes, allowing us to focus on the core aspects of the problem and sidestep any restrictions or biases that might emerge if we were to create more polished models.

Integrated into the modern research process, rapid prototyping plays a vital role in minimizing risk and identifying the feasibility of ideas before resources and time are irreversibly committed. By making use of today's technological advancements, such as 3D printing or digital simulations, researchers can quickly bring abstract concepts to life and assess the impact of proposed solutions in a tangible manner. This approach not only accelerates the research process but also fosters a culture of innovation, where ideas are continually challenged, tested, and improved, pushing the boundaries of our collective knowledge.

In the same way that Tesla or Einstein utilized visual thinking to break the mold and unveil revolutionary insights, this practice has the potential to drive modern research to new heights. By engaging in rapid prototyping, researchers across disciplines can overcome cognitive biases, generate provocative hypotheses, and develop novel solutions to the world's most pressing challenges.

As we venture further into the realms of the visionary mind, let us continue to explore these powerful mental models while keeping in mind their rich historical context and their potential to shape the trajectory of human progress. Armed with a greater understanding of visual thinking and its critical role in innovation, we can transcend the limitations of the present and propel our research to uncharted territories, ensuring a brighter and more profound future for generations to come.

Techniques for Sketching and Mapping Ideas: Creating a Visual Representation of the Problem

Sketching can be the initial step in defining and refining the problem you're attempting to solve. It is an essential tool for not only capturing fleeting ideas but also organizing them coherently. As Thomas Edison once said, "To invent, you need a good imagination and a pile of junk." Sketching is an excellent way to transform that mental "junk" into an organized structure that can fuel discovery. While sketching, you need not worry about artistic skill or aesthetic appearance, as the focus is solely on clarity and comprehension.

First, it is crucial to break the problem down into its core components. Identify the key elements and variables and represent them with simple shapes, such as circles or boxes. Next, use lines or arrows to indicate relationships, dependencies, and interactions between these components. Take Charles Darwin's breakthrough work in evolutionary biology: he represented his insights into the evolution of species in a simple but powerful sketch that led him to the concept of natural selection. Famously known as the "I think" sketch, this visual representation laid out the tree of life structure that would forever change the landscape of biology.

Another powerful technique for creating a visual representation of research problems is mind mapping. Created by Tony Buzan, a mind map is an intricate web of interconnected nodes, with each node representing an idea, concept, or element. It starts with a central theme or question and extends outwards through branches that illustrate relationships between different aspects of the problem. The main advantage of mind maps is their nonlinear structure, which allows for creativity and flexibility in the organization of ideas. Moreover, mind maps leverage the brain's inherent affinity for images, making it easier to associate new information with the existing knowledge base.

To create an effective mind map, begin with a blank canvas, or use a digital tool suited to this purpose. Place the research problem or central theme in the center and let your thoughts flow outwards, exploring related ideas and concepts. Represent each idea with a word, phrase, or image and connect it to the central theme with a branch. Continue adding branches for related ideas or sub-categories, ensuring that you stay relevant to the central theme. Remember, effective mind mapping embraces the organic growth of ideas rather than forcing them into a rigid hierarchy.

Consider the case of Leonardo da Vinci, who is often celebrated for his polymathic genius and incredible innovations across disciplines. When analyzing da Vinci's notebooks, it becomes clear that his extensive use of visual thinking in the form of sketches, diagrams, and mind maps played a crucial role in his groundbreaking discoveries. From exploring the human anatomy to designing intricate machinery, da Vinci's use of visual representation allowed him to dissect complex concepts, reveal unseen connections, and synthesize knowledge from diverse fields.

Once you have practiced these techniques and developed your unique form of visual representation, you can integrate them into your research process. Whether it's identifying gaps in existing knowledge, brainstorming novel solutions, or communicating your results to peers and the public, the power of visual thinking is an invaluable asset for any researcher. As you continue to sketch and map your ideas, always remember the words of the visionary artist Vincent Van Gogh: "I dream of painting, and then I paint my dream." In the realm of research, dreams, and imagination are crucial for the birth of new ideas and innovations.

As you embark on your journey into the world of research, don't forget to express your innermost thoughts through sketching, mapping, and visualization. Channel the interdisciplinary spirit of da Vinci, the tree of life insight of Darwin, or the inventive persistence of Edison. These techniques offer you the opportunity to not only better understand your research but to create a visual language that transcends the limitations of words, facilitates collaboration, and drives groundbreaking discoveries.

Rapid Prototyping Methods: 2D Sketches, 3D Models, and Digital Simulations for Efficient Experimentation

Rapid prototyping methods have become an essential part of the innovation process, allowing researchers to quickly explore and iterate on their ideas before committing extensive resources to the experiments. The practice of rapidly creating physical or digital representations of one's idea is not new, tracing its roots back to Leonardo da Vinci's groundbreaking sketches and Thomas Edison's numerous prototypes. The ability to quickly visualize and test concepts has proven to be invaluable in discovering new knowledge, and with the advent of 2D sketches, 3D models, and digital simulations, the practice took a massive leap forward.

2D sketches are the earliest and simplest form of rapid prototyping. This method involves creating a rough drawing of a concept, allowing the thinker to visualize its various components and relationships. For instance, Nikola Tesla famously used simple sketches to help refine his revolutionary designs for the alternating current motor and transformer. Sketching, as an essential part of the experimental process, allows researchers to make quick adaptations to their designs and readily identify potential flaws or inconsistencies. Moreover, 2D sketches can be shared with colleagues to garner feedback and collaboration, further expediting the ideation process.

3D models, a more advanced form of rapid prototyping, enable researchers to create physical prototypes of their designs. By working with tangible materials, researchers can ascertain the viability of their ideas and evaluate their hypotheses in real-time. This hands-on method often leads to deeper insights and discoveries. For example, James Watson and Francis Crick's landmark 1953 discovery of the DNA double helix structure was accelerated by using physical cutouts of the building blocks and assembling them into different configurations. Working with 3D models offers numerous benefits, such as the ability to test various iterations of designs and examine their functional properties (fit, form, and function) in a tangible manner that is not possible with 2D sketches.

Today, digital simulations further enhance the rapid prototyping process. These virtual models provide researchers with an even greater level of detail, accuracy, and ease in their experimentation. Advanced computer simulations allow researchers to manipulate their designs and visualize complex systems in ways that were previously unimaginable. For instance, cognitive neuroscientist Dr. John Krakauer, creating a digital simulation of a wildly unbeknown system, a stroke recovery paradigm, provided crucial insights into the rewiring process of the human brain.

Moreover, the power of digital simulations is enhanced by the integration of technologies, such as artificial intelligence (AI) and machine learning algorithms, which can help assess and optimize designs more effectively. Researchers can now simulate full-scale experiments in the virtual realm, substantially reducing costs, time, and resource constraints. The pioneering work of physicist Julian Schwinger in developing the first digital simulation of subatomic particles has undoubtedly laid the foundation for today's research in nanotechnology and quantum computing.

Incorporating rapid prototyping methods into scientific research offers researchers a significant advantage as they pursue groundbreaking discoveries. By using 2D sketches, 3D models, and digital simulations to quickly test and refine ideas, thinkers are able to traverse the vast landscape of possibilities and identify the most promising avenues of research more efficiently. This iterative process, grounded in experimentation and rapid realization of mental constructs, allows researchers to unravel the complexities of the natural world and propel the boundaries of human knowledge ever forward.

As we journey deeper into the uncharted realms of scientific exploration, it becomes crucial to continually refine and adapt the tools and techniques at our disposal. The unsolved mysteries and unfulfilled potentials of scientific inquiry beckon us to enhance our research methodologies. By embracing rapid prototyping methods and learning from the ingenuity of history's greatest visionaries, we forge a path toward uncovering the discoveries that lie just beyond the horizon, waiting to change the fabric of our understanding.

Case Studies in Visual Thinking: Examples from Tesla, Einstein, and Other Visionaries

As we continue to delve into the depths of the innovative and creative minds of visionaries, it is essential to examine the role visual thinking plays in their work. Throughout history, some of the most groundbreaking discoveries and monumentally impactful ideas have stemmed from the ability of great thinkers such as Tesla, Einstein, and others to visually conceptualize complex problems and distill them down to their essential components. Through several fascinating case studies, let us explore how these individuals leveraged visual thinking alongside their remarkable intellect to reshape humanity's understanding of the world.

Nikola Tesla, the inventor and engineer behind many revolutionary technologies such as alternating current (AC) electrical systems, was known to manifest his ideas visually within his mind's eye. Tesla firmly believed in the power of visual thinking and often described his mental processes as intense, vivid, and highly detailed experiments that took place within his imagination. Equipped with an eidetic memory, Tesla was reported to be able to construct entire devices and systems mentally before ever putting pen to paper or picking up a tool. This unique capability allowed him to test different iterations, identify flaws, and refine his designs at breakneck speed, all within the confines of his brilliant mind.

For instance, when Tesla was developing his groundbreaking induction motor, he did not resort to traditional drafting methods or mathematical calculations. He envisioned the motor's intricate components rotating in synchronization in his mind before actually constructing the physical prototype. Tesla's ability to test and refine his designs mentally allowed him to build a functional prototype of his induction motor in record time. This example highlights Tesla's extraordinary capacity for visual thinking and his ability to translate a mental vision into a tangible reality that could be shared with the world.

Albert Einstein, perhaps the most famous scientist of all time, was another visionary who harnessed the power of visual thinking to develop groundbreaking theories that radically altered humanity's understanding of the universe. Renowned for his theory of general relativity and his work in theoretical physics, Einstein often relied on thought experiments or "Gedankenexperiments" as a means for engaging with complex scientific problems. Here too, visual thinking played a dominant role in his ability to develop original ideas and groundbreaking theories.

One of the most infamous and cited examples of Einstein's visual thinking is his conceptualization of the "light clock" thought experiment. Proposing a hypothetical scenario where two light beams bounce between two mirrors in a moving and stationary reference frame, the young physicist envisioned the effects of time dilation as dictated by his special theory of relativity. Through his mental visualizations, Einstein revealed that time slows down relative to an observer's speed, a concept that was nothing short of revolutionary during his time. By employing visual thinking to arrive at unconventional conclusions, Einstein cemented his status as one of the most important scientific minds in history.

These anecdotes involving Tesla and Einstein demonstrate the profound potential that visual thinking holds as a conduit for truly groundbreaking concepts and research methods. Like these legendary visionaries, other influential thinkers such as Leonardo da Vinci and Thomas Edison used visual thinking to imagine, design, and invent some of the most iconic creations known to man. Da Vinci's intricate anatomical sketches and Edison's detailed drawings of his numerous inventions serve as evidence of the essential role that visual thinking plays in human progress.

Such case studies illuminate how visual thinking is not merely a tool, but rather an integral aspect of the wiring of exceptional individuals, allowing them to approach problems unconventionally and generate innovative solutions. The immense power of visual thinking lies in its ability to transcend traditional cognitive boundaries, dismantle assumptions, and uncover hidden connections between seemingly unrelated phenomena. By doing so, it becomes an indispensable element in the arsenal of the visionary.

In conclusion, as we aim to cultivate our own visual thinking abilities and implement them into our research methodologies, let us remember the extraordinary impact that visual thinking had on the lives and legacies of Tesla, Einstein, and other pioneering thinkers throughout history. May their stories inspire us to harness the power of visual thinking in our quest for groundbreaking discoveries, forging new terrain across the scientific landscape. For, as Picasso once said, "Everything you can imagine is real." So too shall we dare to imagine and bring our own visions to life.

Integrating Visual Thinking into Your Research Process: Tips and Practical Exercises for Success

One of the most effective ways to start implementing visual thinking into your research is through sketchnoting. Sketchnoting, also known as visual note - taking, is the practice of using simple illustrations, symbols, and arrows to represent ideas or information while taking notes. Instead of solely relying on linear text, sketchnoting allows you to make connections between concepts, identify patterns, and recognize relationships that might not be evident through traditional note - taking methods.

To begin sketchnoting, start by incorporating simple shapes and symbols into your notes to represent the major concepts or themes you encounter in your research. For instance, you may choose to represent the concept of gravity with a drawing of an apple falling from a tree, or the process of photosynthesis with a simple sketch of a leaf undergoing a chemical reaction. This practice will allow you to engage more of your cognitive abilities, leading to a deeper and more thorough understanding of the subject matter.

Another essential aspect of integrating visual thinking into your research is through mind mapping. Mind maps are visual representations of interconnected information, typically organized around a central concept or idea. To create a mind map, start by writing the main topic or concept in the center of the page, then branch outwards from this central idea, adding related subtopics, themes, or questions that emerge from your research. Be sure to use vivid colors, symbols, and images to illustrate your ideas, as this will enhance your brain's ability to process and retain the information.

Mind maps can be especially helpful in brainstorming sessions, generating new ideas, or in the early stages of a research project, as they allow you to visually explore the various facets of a topic and identify potential avenues for further investigation. In addition to helping you organize your thoughts, mind maps also serve as powerful mnemonic devices, making the information you represent more likely to stick in your memory.

Visualizing data is another crucial aspect of visual thinking. Incorporating charts, graphs, or other types of data visualizations into your research can help you better understand complex data sets, identify patterns and trends, and effectively communicate your findings to others. There are numerous tools available, ranging from simple spreadsheet software to more advanced data visualization platforms, that can assist in transforming raw data into visually appealing and informative representations.

One powerful technique used by visionaries like Tesla and Einstein is that of mental simulation, or the ability to mentally test an idea or concept before actually implementing it in the physical world. Try simulating your research problem in your mind's eye by visualizing possible solutions and anticipating potential outcomes. This mental exercise can help you refine your ideas, identify unforeseen obstacles, and make more informed decisions about which approaches to pursue.

In conclusion, the use of visual thinking in the research process arms you with an arsenal of tools and techniques that can drive creativity and enhance your understanding. By incorporating sketchnotes, mind maps, data visualization, and mental simulation into your research, you will be following in the footsteps of the world's most renowned visionaries like Tesla and Einstein. As you continue to leverage visual thinking to advance your own breakthroughs, you will not only deepen your knowledge and improve your problem-solving skills, but you will also be laying the very groundwork for future innovations that may shape the course of human history.

Chapter 4

Harnessing Polymathy: Linking Disparate Fields for Interdisciplinary Breakthroughs

The exploration of knowledge has always faced the challenge of spanning disparate fields to create breakthroughs that revolutionize our understanding of the world. From the polymathic works of Leonardo da Vinci, who intertwined painting, architecture, biology, the arts, and sciences, to Albert Einstein's impressive grasp of multiple disciplines and methodologies, history reveals that some of the most astonishing strides in human progress stemmed from the remarkable ability to see connections between seemingly unrelated areas. Harnessing the energy of polymathy-the blending of diverse domains-offers the potential to unlock insights and create interdisciplinary breakthroughs that drive the frontiers of research into new and exciting territories.

One example of this strength unleashed by polymathy can be seen through the work of notable physicist, Richard Feynman. While primarily known for his revolutionary work in quantum mechanics, Feynman's curiosity-driven endeavors in a multitude of fields led him to discover groundbreaking connections that ultimately contributed to unexpected advancements in subjects as varied as artificial intelligence and nanotechnology. By synthesizing elements from different areas, Feynman developed original insights on topics such as: quantum electrodynamics, weak-force theory, lattice quantum chromodynamics, and more. As a result of his insatiable drive to learn, experiment, and push traditional boundaries, these interdisciplinary connections made by Feynman transformed the fields of study that he touched.

However, emulating a polymathic approach towards problem-solving is often easier said than done. Society tends to reward specialization, placing more value on a focused and strategic pursuit of knowledge within a niche. This perception often pushes researchers to dive deep within a discipline, hoping to solve complex problems by drilling down to the most fundamental principles of their topics. Acknowledging these structural dynamics, fostering polymathy requires overcoming the innate predisposition towards compartmentalized thinking.

In cultivating interdisciplinary thinking, researchers can expand their intellectual horizons by acquiring foundational competencies in diverse fields. Building this diverse base of knowledge allows them to see connections and generate novel ideas by engaging in convergent thinking. Convergent thinking gives voice to the extraordinary potential for innovation by traversing disciplinary boundaries and constructing unique insights from a broader perspective. This more expansive understanding can then be utilized for investigating and developing new hypotheses and theories that bring discrete fields of study together, thereby prompting a new iteration of groundbreaking research.

For a practical example of this, consider the field of material science. This seemingly narrow domain involves the synthesis of discoveries from physics, chemistry, mathematics, and engineering, boasting innovations such as Kevlar, superconductors, and revolutionary microprocessors. Furthermore, by uniting knowledge in fields as diverse as biology and technology, researchers have built artificial muscles, and other bio-inspired technologies are on the horizon. These remarkable advances shine light upon the hidden possibilities that lie undiscovered at the intersection of various disciplines.

Another example can be seen in the development of modern artificial intelligence (AI), which utilizes game theory, computer science, neuroscience, and psychology to solve problems. AI breakthroughs have led to profound improvements in areas like personalized medicine and environmental conservation. The technology and insights garnered from the merging of these diverse fields have driven the rapid progress which gave birth to autonomous vehicles and quantum computing.

In traversing the uncharted spaces between formal disciplines, polymathy is the force that connects the seemingly unconnectable. It is what weaves the intricate tapestry of human knowledge, providing shape and cohesion to our understanding of the universe and ourselves. By embracing and cultivating interdisciplinary thinking, researchers can gain the tools needed to invent novel ideas, discover new connections, and push the frontiers of modern research into unanticipated and innovative realms.

In the quest to propel the journey of problem - solving, we must acknowledge the limitations of diving deep into a singular discipline. We must recognize the fantastic power of a polymathic perspective in encompassing the world's mysteries from a panoramic vantage point. For when the winds of inspiration blow across these intersecting landscapes, the true wonder of imagination comes alive with the spark of breakthroughs that have the potential to redefine history. Ultimately, it is at the nexus of disparate realms that the extraordinary language of innovation speaks most profoundly, carrying the markers of progress across the boundless continuum of human understanding.

The Power of Polymathy: An Introduction to Interdisciplinary Thinking

A crucial aspect of polymathy is the propensity to transcend the confines of one's primary field of expertise and draw inspiration from diverse fields. Such intellectual versatility allows polymaths to integrate knowledge from different domains and identify unique patterns, relationships, and possibilities. A classic example of polymathy can be found in Leonardo da Vinci's insatiable curiosity and diverse interests, which spanned from anatomy and botany to engineering and painting. This diversity of interests allowed him to discover connections that eluded his contemporaries, leading to remarkable innovations such as the concept of the parachute and the Vitruvian Man.

Moreover, polymathy transcends the mere accumulation of knowledge; it is a multidimensional endeavor that nurtures creative problem - solving, synthesis, and adaptability. By continually broadening their intellectual horizons, polymaths cultivate an eagle - eyed vision that affords them the unique ability to perceive and elucidate novel connections between disciplines. This synthesis of ideas is superbly exemplified by Ada Lovelace, whom many regard as the first computer programmer. Lovelace not only envisioned an early model of a computer, but her background in mathematics and her artistic sensibilities allowed her to explore the possibility of the machine's potential recreation of not just numbers but complex symbols and art.

Polymathy also necessitates a willingness to challenge existing paradigms and shatter the echo chambers that often stifle innovation within insulated disciplines. By deftly navigating the terrain of different fields, polymaths cultivate intellectual resilience and an antifragile mindset - one that sees value in uncertainty and the unknown. Their intrepid explorations enable them to question entrenched assumptions, embrace dissenting perspectives, and synthesize diverse schools of thought. Benjamin Franklin serves as an eloquent representation of such intellectual fearlessness: a writer, scientist, politician, and inventor whose endeavors reshaped our understanding of electricity and provided the foundation for revolutionary inventions such as the lightning rod and bifocals.

In an era where hyper - specialization often predominates, cultivating polymathy is more crucial than ever. The complex, interdisciplinary nature of contemporary problems demands an approach that transcends the myopic lens of a single discipline. Tackling grand challenges such as climate change, artificial intelligence, and socio - economic inequality requires a mental toolkit that can traverse the boundaries of traditional expertise and integrate insights from disparate sources. Only by fostering intellectual diversity and openness to new ideas can we hope to unravel the Gordian knots that bind our collective progress.

To pave the pathway towards polymathy, individuals must be relentless in their pursuit of knowledge, even beyond their chosen field, embracing the unfamiliar and unknown. Imbibing a growth mindset, nurturing intellectual humility, and engaging in continuous learning can help strengthen the diverse cognitive muscles that underpin polymathy. Practical techniques for cultivating polymathy include the deliberate synthesis of unrelated disciplines through personal projects, artist dates, reverse mentoring, and interdisciplinary book clubs.

As we forge ahead into an increasingly complex and interconnected world, polymathy will play a pivotal role in shaping the contours of human knowledge and innovation. By embracing the power of interdisciplinary thinking, we not only expand the repository of human wisdom but tap into the latent creative potential that lies at the intersection of disparate fields. As such, the polymathic mindset must be nurtured not only in individuals but across scientific, academic, and professional institutions at large. Only then can we construct a future where collaboration and ingenuity serve as the bedrock for unprecedented breakthroughs.

In the annals of human achievement, the polymath stands as a testament to the boundless potential of interdisciplinary thinking. As we venture forth into the uncharted territories of research and discovery, let us remember the visionary legacy of those who dared to traverse the borders of intellectual realms, paving the way for future generations to dare, dream, and explore, unbounded by the limitations of disciplinary confines.

Historical Examples of Polymaths: Da Vinci, Benjamin Franklin, and Ada Lovelace

Throughout history, the minds of polymaths have stood as living testaments to the power of interdisciplinary thinking and the limitless bounds of human curiosity. Leonardo da Vinci, Benjamin Franklin, and Ada Lovelace, each a polymath in their own right, exemplify the diverse and unique gifts that come from exploring multiple disciplines. The biographies of these great thinkers reveal important insights about cultivating a broad foundation of knowledge. By examining their lives and the various fields of study they pursued, we can begin to unravel the mystery underlying their intellectual prowess, and understand how polymathy contributed to their unparalleled success and groundbreaking contributions.

Leonardo da Vinci, a towering figure of the Italian Renaissance, remains one of the most recognizable and celebrated polymaths. He sought knowledge in a multitude of areas, including anatomy, engineering, optics, mechanics, and painting. Leonardo's boundless curiosity led to a profound understanding of the natural world, which he applied to a wide range of disciplines. For example, his anatomical renderings of the human body, such as the Vitruvian Man, seamlessly blend his mastery of artistic technique with his deep understanding of human physiology. By deftly navigating his way through the boundaries that separated disciplines, Leonardo was able to establish a holistic understanding of the world, which allowed him to forge new connections and uncover principles that were hitherto unknown.

Similarly, Benjamin Franklin, the American polymath, was an author, entrepreneur, scientist, inventor, and statesman. His diverse range of interests included electricity, meteorology, navigation, printing, and politics. Recognizing common properties between electricity and lightning, Franklin carried out a series of experiments, culminating in the invention of the lightning rod, a life-saving device that is still in use today. His interdisciplinary approach also enabled him to design bifocal eyeglasses by combining optical knowledge with elements from both convex and concave lenses, solving a fundamental problem faced by people with varying degrees of visual acuity. Franklin's natural inclination to seek knowledge in multiple disciplines and his ability to link disparate ideas to form innovative solutions was a crucial determinant of his substantial contributions to humanity.

Ada Lovelace, the Countess of Lovelace and daughter of the famous poet Lord Byron, found her calling in the emerging world of computer science. Primarily remembered as a mathematician and writer, Lovelace is often regarded as the world's first computer programmer. What set Lovelace apart from other mathematicians of her time was her capacity to apply her mathematical skills to an abstract concept - Charles Babbage's Analytical Engine, a precursor to modern computing devices. As Babbage struggled to secure funding for his invention, it was Lovelace who had the vision to see the potential of the machine, recognizing that it could process not only numbers but symbols and even music. Her willingness to venture into a new field and develop original ideas reflected the essence of polymathy and its ability to facilitate groundbreaking discoveries.

In examining the lives and works of these incredible historical polymaths, we find a common thread of insatiable curiosity, which guided their pursuit of knowledge across numerous domains. Leonardo da Vinci, Benjamin Franklin, and Ada Lovelace each challenged the boundaries set by their contemporaries, integrating knowledge from disparate disciplines to produce revolutionary advances that would reshape their respective fields. The trinity of these polymaths exemplifies the power of interdisciplinary learning and the influence of creativity and curiosity on scientific breakthroughs.

As we forge ahead into the depths of the modern research landscape, burdened by the weight of specialization, the examples set forth by these polymaths serve as a beacon, igniting the sparks of curiosity and illuminating the possibility of a holistic intellectual experience. Polymathy is not merely a relic of the past; it is a call to action for those of us who wish to unleash the untapped potentials of the human mind. By embracing the lessons gleaned from these historical figures, we might yet unlock the doors to future innovations and propel humanity toward new horizons of understanding.

Practical Techniques for Cultivating Polymathy: Building a Foundation of Broad Knowledge

In the pursuit of scientific breakthroughs and the furthering of human understanding, cultivating polymathy - the knowledge and mastery of multiple fields of study - can be an invaluable asset. However, the task of building a foundation of broad knowledge can seem daunting, especially when our modern education system often encourages specialization. Let's delve into some practical techniques you can employ to embrace and foster the polymath within you, unlocking the potential for innovation and discovery that comes with interdisciplinary thinking.

One of the most effective strategies for nurturing polymathy is to develop a routine of continuous learning. Although it may seem obvious, dedicating some time from each day to engage in different fields of study is the first step in expanding your intellectual horizons. Reading books, articles, and academic papers encompassing a wide range of subjects can help you familiarize yourself with the foundational theories, methodologies, and research questions that define each discipline. By dedicating a consistent amount of time to acquiring new knowledge routinely, you will gradually build a strong interdisciplinary understanding that naturally leads to connections between fields.

Equally important as learning across different fields is engaging in active learning experiences. Hands - on activities, such as working on projects, volunteering, or attending workshops within various domains, can greatly enhance your ability to integrate and apply what you have learned. Practical experience will enable you to establish a deeper comprehension of the field, and may also spark your creativity to approach problems from unique perspectives.

Another technique for developing polymathy introduces aspects of deliberate practice - the act of breaking down skills and knowledge into smaller pieces and mastering them individually. Understanding the essential components of each field and their relationships can help you create a mental architecture upon which you can build a broader contextual understanding. Be persistent in revisiting and refining your mental frameworks, as they form the foundation for pattern recognition, creative thinking, and interdisciplinary connections.

As a budding polymath, don't be afraid to dive into challenging material in the fields you explore, even if it's beyond your current level of expertise. Tackling difficult subject matter is essential for intellectual growth and fundamental understanding. Conversely, every subject has its jargon and terminology, which can become a barrier if not appropriately addressed. Having a glossary handy while diving into such material can aid in understanding its inherent intricacies and reduce mental friction.

One of the defining aspects of polymaths is their ability to see connections and draw inspiration from seemingly disparate areas of knowledge. Developing this skill requires the deliberate synthesis of information from different domains. This can be practiced by keeping a journal or blog, where you can jot down insights, interesting facts, or unanswered questions from various fields and periodically reviewing them to find potential connections and patterns. This process helps train your brain to identify links and associations between related fields, expanding the realm of possibility for innovative ideas.

Critical engagement with different intellectual communities is another effective tool when cultivating polymathy. Conversations with experts, attending conferences and seminars, and joining online forums in diverse fields can sharpen your thinking and expose you to ideas that might not have crossed your path otherwise. Engaging with distinct intellectual communities often illuminates the common struggles and the fundamental concepts that resonate across disciplines.

As you immerse yourself in multiple fields, it's important to keep in mind the principle of intellectual humility: a recognition of the limitations of your knowledge. This attitude enables you to approach new subjects with curiosity and an openness to learning, fostering genuine growth and deeper understanding. Additionally, fostering your creativity through artistic expressions such as painting, writing, or playing an instrument can complement your newfound polymathy to unleash an immense creative potential.

In the spirit of intellectual exploration, let our voyage into polymathy embrace the words of famed poet William Blake, "To see a world in a grain of sand, and heaven in a wildflower." With dedication, curiosity, and persistence, you can cultivate the polymath within you, expanding your horizons and becoming a harbinger of interdisciplinary breakthroughs yet unknown. As we dive into the connections between biology, physics, art, and beyond, remember that the tapestry of human knowledge is intricately interwoven, and every discipline holds the potential to inform and inspire another.

Identifying Patterns and Connections Across Disciplines: Linking Biology, Physics, Art, and More

A sudden realization overcame the young Isaac Newton as he sat quietly beneath the apple tree. Watching the apple plummet towards Earth, Newton's mind raced as he understood gravity's invisible pull on objects in a whole new light. Like other polymaths, Newton had the unique ability to identify connections across disparate disciplines - in this case, mathematics, physics, and astronomy - which would revolutionize scientific thinking and forever alter the course of history.

Historically, where visionary thinkers have triumphed in the advancement of human knowledge is precisely in these moments, wherein seemingly unrelated disciplines collide, providing opportunities for new insights and groundbreaking hypotheses to emerge. We find that, time and again, the most profound discoveries are derived from the exploration of these intersections, which yield hybrid theories, paradigms, and technologies.

Drawing clear connections between distant fields, such as biology and physics or art and mathematics, may appear counterintuitive at first glance. However, the rewards of investigating their shared principles and underlying mechanisms can be both inspiring and transformative.

Take, for example, the study of protein folding, which straddles the border between chemistry and biology with great finesse. Scientists in this field have discovered that certain protein structures, the building blocks our cells, rely on principles of energy minimization which are eerily similar to those found in physics. Just like water always takes the path of least resistance when flowing downhill, proteins 'fold' into their most stable conformations, guided by the natural proclivity toward the lowest energy state. Recognizing this common denominator between the physical and biological world has allowed researchers to approach the mystery of protein folding in new and intricate ways, furthering our understanding of this vital cellular process.

Similarly, artistic visions have often influenced scientific innovations and vice versa. The Italian polymath Leonardo da Vinci provides a prime example of the intermingling between art and science. His study of anatomy and in-depth exploration of human form revealed insights into the human body that challenged traditional medical thinking, imbuing his own artistic renderings with novel perspectives. Perhaps even more compelling is da Vinci's fascination with flight, his mind linearly drawing on the airborne mechanics of bat wings and the winding path of a flying bird. These artful investigations ultimately culminated into detailed, technical sketches for various contraptions, including the ornithopter and the aerial screw, which foreshadowed modern-day helicopters and airplanes.

In each of these cases, the synchronicity of disparate fields has been a catalyst for innovation and creation. Thus, to further develop one's polymathic thinking, the first challenge is to identify these unobvious connections, patterns, and relationships between what seemingly lies far apart. Subsequent to such a realization, one must embrace curiosity and explore these intersections with an open mind. This requires the suspension of our typical tendency to compartmentalize knowledge, and instead, view the world through an interdisciplinary lens. Challenges and skepticism will inevitably arise, but perseverance and resilience will yield a wealth of novel insights.

Finally, in order to continually cultivate these cross-disciplinary connections, one must actively seek the integration of diverse ideas and fields and encourage collaboration with experts from varied backgrounds. In building a proverbial bridge between the disparate fields of thought, we create a common ground for leaps of understanding and boundless possibilities.

The journey of interdisciplinary thinking is a rousing one, as it enlarges the boundaries of traditional disciplines and beckons us into uncharted territory. In doing so, polymathic thinking illuminates the obscure and plots a captivating course to future innovations and discoveries. As we embark on this voyage, let us heed the wisdom of Albert Einstein, who famously said, "The only source of knowledge is experience." Embrace the pursuit of knowledge not in isolation or rigid segments but in a grand, interconnected tapestry of wisdom that unites us all.

The Role of Polymathy in Future Innovations: Cooperation Between Different Fields of Research

The rise of innovation in recent decades has been a testament to the ever-increasing human capabilities, fueled in part by the cooperation between different fields of research. This transformative progression can be predominantly attributed to the polymathic nature of modern scientific and technological breakthroughs. Polymaths, who are characterized by their mastery of multiple disciplines, have long played a significant role in advancing human knowledge, enabling them to build bridges between seemingly unrelated subjects, facilitating their mutual enrichment.

A striking example of polymathy in action is encapsulated in the application of game theory to biology, leading to the establishment of a new field known as evolutionary game theory. This branch of biology focuses on the interaction between individual organisms, which is of paramount importance in determining the evolution of populations. Stemming from mathematical models used to predict the best strategy in competitive situations, game theory found its way into biology, notably, through the works of polymaths like John Maynard Smith and George Price. Ultimately, the interdisciplinary application of game theory to biology exponentially enhanced our understanding of how natural selection drives evolution, illustrating how synergistic cooperation at the crossroads of knowledge can propel us into uncharted territories of innovation.

Furthermore, the advent of artificial intelligence (AI) systems owes much of its existence to the intersection of various fields such as computer science, neuroscience, psychology and philosophy. The development of AI requires deep understanding and collaboration between researchers from these diverse domains and, unsurprisingly, many of the pioneers in AI, such as Alan Turing and Marvin Minsky, are themselves acknowledged as polymaths. The use of AI continues to permeate our lives in multiple ways, from virtual assistants to predicting stock market trends, and consistently drives technological advancement across disciplines.

The future trajectory of innovation heavily relies on the ability of researchers to not only have an acute awareness of their own field but also possess the intellectual adaptability necessary to approach problems from alternative perspectives, drawing inspiration from seemingly unrelated domains. As data generation and dispersal accelerates unabated in the age of information, the capacity for polymathic thinking to synthesize information across disciplines become increasingly vital.

A prominent example of polymathic innovation can be found in the realm of biomimicry; where hundreds of novel solutions have been devised by mimicking the exceptional and diverse mechanisms observed in nature. For instance, the aviation industry has benefited from the interdisciplinary study of bird flight mechanics, leading to design improvements and increased fuel efficiency. Similarly, advancements in robotics are continuously inspired by studying the locomotion and physiology of animals, which paves the way for developing innovative designs. In such cases, breakthroughs in engineering and technology draw upon expertise in multiple fields, including biology, material sciences, and design, demonstrating a holistic approach to problem - solving.

It is increasingly evident that the future of innovation will continue to demand multidisciplinary collaboration. An essential aspect of this strategy is fostering a culture of mutual respect and curiosity amongst researchers from differing fields. Creating spaces that stimulate and nurture interdisciplinary cooperation, such as research institutes with diverse intellectual portfolios and academic conferences that bring together experts from varied domains, adds to the potential for uncovering new insights and connections.

Moreover, integrating polymathic thinking into education systems can nurture the next generation of innovators and researchers equipped with the tools to generate cutting - edge ideas across disciplines. Encouraging the youth to embrace a polymath mindset could potentially lead to an unforeseen cascade of scientific advancements.

As we embark on this ever - evolving journey of discovery, it becomes increasingly imperative to look beyond the paradigmatic boundaries that delineate traditional fields of research, fostering an environment that encourages transdisciplinary connections. To unlock the colossal potential for groundbreaking discoveries, we must continue to navigate this rich and intricate tapestry of human knowledge, intertwining threads of diverse intellectual talent that underpin the colorful masterpiece of innovation. Fortifying this ecosystem of polymathy and interdisciplinary cooperation will ultimately pave the way for realizing our collective aspirations, propelling humanity onto a boundless trajectory of unprecedented insights and breakthroughs.

Challenges and Benefits of Interdisciplinary Collaboration: Overcoming Barriers and Embracing Synergy

History is replete with instances of fruitful interdisciplinary collaboration. Consider the Human Genome Project (HGP), where the collaboration between biologists, mathematicians, computer scientists, and engineers led to a revolutionary understanding of our genes and their complex interactions. The HGP fostered the development of computational biology, bringing new techniques for sequencing, data analysis, and modeling genetic phenomena. Such a breakthrough would not have been possible without the joint endeavor of experts from different fields working in concert.

Another prime example is the discovery of graphene, a novel material with a myriad of applications in electronics, materials science, and energy storage. Two physicists at the University of Manchester, Andre Geim and Konstantin Novoselov, stumbled upon graphene when they used a piece of adhesive tape to lift a layer of carbon atoms from a chunk of graphite. Their subsequent collaboration with chemists, materials scientists, and engineers opened up new avenues of research, culminating in the 2010 Nobel Prize in Physics for their pioneering work.

While interdisciplinary collaboration can result in revolutionary breakthroughs, it is not without challenges. One major hurdle lies in bridging the gap between different academic languages and methodologies. Jargon and technical tools from separate disciplines can be alienating, leading to confusion and miscommunication. To overcome this barrier, collaborators must invest time and effort in understanding and appreciating each other's field, developing a common language, and working to translate their individual insights into a collective understanding of the problem at hand.

A related challenge is the resistance to change within established academic cultures. Institutions are often structured around disciplinary boundaries, limiting the opportunities for cross-disciplinary interactions. Moreover, training and education within specific disciplines can engender a certain level of dogmatism and rigidity in thought, hindering the ability to visualize problems from multiple perspectives. To address this challenge, fostering a culture of curiosity and open-mindedness, both within and beyond the confines of academic institutions, is essential.

A successful interdisciplinary collaboration should embrace the concept of synergy, the idea that the whole is greater than the sum of its parts. Capitalizing on the unique strengths of each collaborator while maintaining a united intellectual vision is paramount. Achieving this balance requires humility, empathy, and ongoing communication among team members. Regularly exchanging ideas, asking questions, and providing constructive feedback can propel the group forward in its shared journey of discovery.

Instituting dedicated interdisciplinary programs and conferences, as well as encouraging collaboration across departments within institutions, can facilitate the mutually beneficial exchange of insights, leading to remarkable discoveries. For example, the Santa Fe Institute in New Mexico fosters interdisciplinary research, inviting experts from various disciplines to conduct research on complex systems, a field where diverse intellectual traditions readily coalesce.

As we venture into a future characterized by increasing complexity and interconnectedness, the power of interdisciplinary collaboration becomes increasingly pertinent. Drawing from the experiences of the past and embracing the challenges of the present, we must foster a global research community that champions curiosity, collaboration, and the unbound potential of diverse perspectives. By overcoming the barriers to interdisciplinary collaboration and embracing the synergy that emerges from the intersection of disciplines, we stand to unlock the next wave of groundbreaking innovations that will reshape the course of human history.

Chapter 5

Overcoming Cognitive Biases: Cultivating Objectivity and Beginner's Mind in Research

Mastering the art of scientific discovery requires not only technical prowess, but also the ability to challenge our own assumptions and free ourselves from the cognitive biases that can constrain our perspective. It is through the cultivation of a beginner's mind, as Zen master Shunryu Suzuki once said, "in the expert's mind, there are few possibilities; in the beginner's mind, there are many," that we can achieve a more objective lens and uncover the true potential of our research.

Renowned physicist Richard Feynman once emphasized the importance of unbiased scientific inquiry, stating, "it doesn't matter how beautiful your theory is, it doesn't matter how smart you are. If it doesn't agree with experiment, it's wrong." In this light, understanding and mitigating the influence of cognitive biases can empower us to become more effective scientists capable of driving research breakthroughs.

Let us revisit the story of Louis Pasteur. While experimenting with bacteria, he formed a hypothesis that would later become the foundation of the germ theory of disease. Competition with another scientist led Pasteur to tap into the creative power of cognitive reframing. He chose to view this rivalry not as an impediment but as an opportunity for growth. This shift in perspective allowed him to scrutinize his own conclusions and dive deeper into his research, culminating in a better understanding of how some diseases are caused by tiny microorganisms. By questioning his existing beliefs and embracing the challenge of reconsidering his theory, Pasteur ultimately transformed the world of medicine.

Cultivating objectivity in research starts by building awareness of our cognitive biases, including but not limited to, confirmation bias, recency bias, and anchoring. To elicit the true potential of groundbreaking ideas, one must first expose these inherent mental shortcuts to scrutiny. While these biases have evolved to ensure rapid and efficient cognitive processing, they can hinder our ability to think beyond existing frameworks and blind us to alternative perspectives.

A particularly effective strategy for overcoming cognitive biases is seeking external input. Collaborating with researchers from different backgrounds and expertise can challenge our assumptions and broaden the scope of our analysis. Taking the time to engage in purposeful discussions and debates enriches the thinking process, enabling us to refine our hypotheses and venture into uncharted territories.

Moreover, through mindfulness practices, we can develop a heightened sense of self-awareness that allows us to recognize and confront our cognitive biases head-on. Deepening our understanding of our personal predispositions fosters mindful scrutiny of our thought processes, facilitating more objective decision - making. By adopting a beginner's mind, we become open to the many discoveries that might have been overlooked by an expert mind clouded by mental models and framework rigidity.

The journey towards cultivating objectivity requires not only diligence but also an unwavering commitment to challenge the pillars of scientific belief. As the story of Pasteur demonstrates, maintaining a posture of intellectual curiosity and embracing the value of collaboration can guide researchers through a labyrinth of uncertainty into the realms of ground-breaking discovery.

As we forge our paths towards becoming visionary researchers, let us embrace the wisdom of Zen Master Shunryu Suzuki, who understood that the true spirit of innovation is a state of perpetual inquiry. By harnessing the power of objectivity and cultivating a beginner's mind, we sharpen our intellects and strike gold from the mines of scientific discovery. It is through this unrelenting quest for the unknown that we secure our place as pioneers in our respective fields, and drive the boundaries of human knowledge into territories yet to be explored.

Introduction to Cognitive Biases: How They Affect Research and Discovery

In the quest for groundbreaking discoveries and new insights, researchers and scientists rely on their capacity for critical thinking and objective judgment to further their respective fields. Yet, our cognitive abilities are not immune to pitfalls and biases which may cloud our decision - making process, potentially leading to false conclusions, or worse, obstructing the path to true understanding. Hence, it is of utmost importance for any intellectual endeavor to recognize the existence and impact of cognitive biases and learn how to navigate through or around them in order to maintain a clear path toward scientific progress.

Cognitive biases are defined as systematic errors in thinking, leading to deviations from objective reality or rational judgment. These biases arise from our brain's heuristic mechanisms, which are designed to simplify the complex world around us and enable us to make quick decisions. However, this very process of simplification and approximation occasionally yields a distorted perspective, which may not always serve us well in the realm of research and discovery.

Take, for example, the infamous case of the discovery of nuclear fusion by electrochemists Martin Fleischmann and Stanley Pons in 1989. The two researchers claimed to have observed cold fusion - a form of nuclear reaction at room temperature - while conducting experiments with palladium electrodes immersed in heavy water. Their announcement sparked intense excitement and hope for a revolutionary new energy source. However, it was later revealed that their findings were likely tainted by confirmation bias, as they selectively focused on a particular set of data that seemed to support their hypothesis while ignoring other contradicting evidence. This led to hasty claims of cold fusion which could not be replicated by other researchers, culminating in the discrediting of this premature and unfounded discovery.

The case of Fleischmann and Pons demonstrates the subtle but powerful influence of cognitive biases on scientific endeavors. Confirmation bias, in particular, is of special interest in the realm of research as it tends to skew our analysis of data and undermines our pursuit of objectivity. Though these biases are a result of our brain's natural decision-making and evaluation processes, they can also be mitigated and controlled through conscious effort and self-awareness, honing our mental tools and sharpening our perception of reality.

Another compelling example of cognitive bias affecting research and discovery is the story of Joseph Jastrow, a psychologist who, in 1899, sought to investigate the relationship between the perception of size and the mental processing of illusions. He conducted a series of experiments which involved showing participants his now-famous "Jastrow illusion" - where two equal-sized shapes appear to be of different sizes due to their relative positioning. However, Jastrow unwittingly fell into the trap of experimental bias, as he had a strong expectation that the participants would misperceive the sizes. This expectation unconsciously prompted him to provide subtle cues or hints that potentially influenced the subjects' responses. Indeed, subsequent independent replications of the experiments - carried out without the possible influence of experimental bias - yielded drastically different results.

The Jastrow illusion case is a stark reminder that even the most meticulous and well-intentioned researcher is susceptible to cognitive biases, which can compromise the validity of their work. The recognition and mitigation of these biases are, therefore, paramount in maintaining the integrity of our intellectual pursuits.

In conclusion, our exploration of the cognitive landscape should echo the words of physicist Richard Feynman, who advocated for the need to "learn from science that you must doubt the experts" and "leave the door to the unknown ajar." Accepting that our cognitive abilities are indeed fallible allows us to confront our biases, refine our methods, and elevate the quality of our research. While the challenges posed by cognitive biases may hinder discoveries and breakthroughs, it is within our power as researchers to cultivate objectivity, adaptability, and humility, propelling our quest for knowledge ever forward, toward a horizon of scientific enlightenment and transformative progress.

Common Cognitive Biases in Research: Confirmation Bias, Recency Bias, and Others

Cognitive biases are systematic deviations in judgment from normative rationality, causing individuals to draw erroneous conclusions based on their prior beliefs or information. These mental shortcuts can have significant implications in various fields, including scientific research. By understanding the most common cognitive biases in research, scientists and researchers can work toward minimizing their impact and ensuring objective and reliable findings.

One of the most prevalent cognitive biases in research is confirmation bias. This bias occurs when an individual seeks out or interprets evidence in a way that confirms their preconceptions, leading to disregard for contrary evidence and possible distortion of findings. For example, a researcher convinced that a specific treatment is effective may be more prone to emphasize positive results and downplay negative ones. This tendency might lead to the belief that the treatment is more effective than it truly is, potentially leading to flawed conclusions. A classic example of confirmation bias can be found in the Cold Fusion controversy, in which researchers Martin Fleischmann and Stanley Pons announced groundbreaking findings in 1989 that they believed demonstrated the possibility of cold fusion. Subsequent attempts to replicate their results failed, but the duo remained steadfast in their belief, dismissing the failures as a result of the experimentation inconsistencies that others introduced.

Another common cognitive bias is recency bias, which occurs when an individual gives disproportionate weight to the most recent information they receive, neglecting or underestimating the importance of earlier data. The danger of recency bias is particularly apparent in research where data is collected over extended periods of time. By giving preference to recent data points, a researcher may overlook trends or vital information present in the earlier data, potentially leading to erroneous conclusions. A notable case of recency bias can be observed in the field of climate science: public opinion on climate change may be temporarily swayed by recent weather events, regardless of the broader trends and the Socratic accumulation of evidence over decades.

Anchoring bias represents another threat to objective research, particularly during the process of data interpretation and hypothesis generation. This bias describes the human tendency to rely heavily on the first piece of information encountered (the "anchor") when making decisions or judgments. For example, researchers may unconsciously anchor their estimation of a variable based on a preliminary piece of data they observed, leading to inaccurate conclusions further down the line. A real - life illustration of anchoring bias can be found in the Challenger shuttle disaster, where pre - launch estimations of the O - ring failure rate were anchored on low probabilities, despite mounting evidence to the contrary.

Availability bias describes the propensity to overestimate the likelihood of an event based on the ease with which instances of that event come to mind. This bias may affect research when formulating hypotheses, as researchers may favor hypotheses that are easily recalled or are top - of - mind, ignoring less available but potentially more plausible explanations. For instance, the disproportionate fear of shark attacks relative to other, more frequent threats can be attributed to the availability bias, as media reports and images of shark attacks are more vivid and easier to recall than reports of other, less dramatic dangers.

By recognizing the common cognitive biases in research - confirmation bias, recency bias, anchoring bias, and availability bias - researchers can take proactive steps to mitigate their effects and ensure a higher degree of objectivity and rigor in their scientific pursuits. This awareness can lead to a more robust research process and pave the way for groundbreaking discoveries that stand up to scrutiny. With an emphasis on revisiting biases frequently and remaining self - aware, scientists can avoid the pitfall of becoming a slave to their cognitive machinery, ensuring that their findings are firmly rooted in data and not distorted by the subconscious reverence for flawed expectations or flashy, facile impressions.

Strategies for Identifying and Mitigating Cognitive Biases in the Research Process

As Carl Gustav Jung aptly stated, "The pendulum of the mind oscillates between sense and nonsense, not between right and wrong." This observation is particularly striking in the field of research. Cognitive biases, which are systematic patterns of deviation from rationality or good judgment in decision making, can negatively impact the scientific process, causing errors in data interpretation, hypothesis formation, and experimental design. However, several strategies can be employed to identify and mitigate these biases, ensuring that research remains rooted in objectivity and truth.

First, it is essential for researchers to develop an awareness of the specific cognitive biases to which they may be susceptible. These biases may vary among individuals and can include confirmation bias (favoring information that supports one's existing beliefs), recency bias (giving undue weight to recent events), and anchoring bias (relying too heavily on an initial piece of information, or "anchor," when making subsequent judgments). By familiarizing oneself with these biases and their potential impact on research, a researcher can be vigilant in spotting them when they arise.

One practical method for identifying cognitive biases is to engage in reflective thinking or journaling throughout the research process. Researchers should record their thought processes, hypotheses, and interpretations of data, then take time to critically challenge their assumptions and beliefs. By engaging in regular self-assessment, researchers can identify instances when they may be unknowingly swayed by their biases and make conscious efforts to counteract them.

Mitigating biases requires a deliberate adjustment in thought patterns and behaviors. Research has shown that cognitive biases can be reduced through a process called "debiasing," which can involve a range of techniques, including perspective-taking, seeking conflicting evidence, and employing statistical reasoning skills. For example, a researcher can adopt a "devil's advocate" approach by actively seeking out evidence that contradicts their current beliefs and then reevaluating their hypotheses in light of the new data.

Another powerful tool to counteract cognitive biases is collaboration with others who possess diverse perspectives, backgrounds, and expertise. Collaborating on research projects can challenge one's assumptions, promote intellectual growth, and introduce new ideas. Furthermore, multidisciplinary groups are less likely to fall victim to groupthink, as each member brings a unique viewpoint and skill set. Additionally, seeking feedback from trusted colleagues or mentors who are familiar with your work can help identify instances when cognitive biases may be influencing your research.

In some cases, cognitive biases may stem from incomplete or inaccurate knowledge of the subject matter. Therefore, researchers should actively engage in continued learning initiatives and remain informed about developments in their field. Attending conferences, seminars, and lectures related to one's research area can provide fresh insights and inspire new ways of thinking, combating cognitive biases by expanding one's knowledge base.

If biases persist despite these measures, researchers may consider turning to technology for assistance. Machine learning algorithms can be trained to detect and correct for biases in data analysis or interpretation. For example, computer models can identify trends and correlations in data, which, when viewed with a human's subjective lens, might be susceptible to cognitive biases. By leveraging such technology, researchers can ensure their results remain grounded in objective analysis.

In the realm of science, the philosopher Arthur Schopenhauer's words resonate with particular force: "Every man takes the limits of his own field of vision for the limits of the world." It is the responsibility of the researcher to overcome these limits by identifying and mitigating cognitive biases and cultivating a sense of intellectual openness and curiosity. By utilizing these strategies, researchers can ensure that their work transcends such limitations, allowing them to unveil the hidden truths that lie beyond the confines of their own mind's pendulum. As we venture deeper into the vast expanse of human knowledge, using these strategies will not only enrich our understanding of the mysteries that envelop us but might also reveal unexpected connections and spark revolutionary discoveries - the kind that have shaped science and human progress throughout history.

Cultivating Objectivity: Techniques for Ensuring an Unbiased Perspective

Objectivity, the practice of making judgments based on facts and evidence rather than emotions or preconceived notions, is a fundamental component of robust research. However, maintaining an unbiased perspective is no easy task, as humans are susceptible to a variety of cognitive distortions. These distortions can manifest themselves in the form of subconscious biases, selective attention, or the natural tendency to validate our own beliefs. By employing a range of techniques, researchers can minimize these influences and cultivate a more objective mindset.

One key strategy is to adopt an attitude of curiosity, which centres around exploring a wide array of perspectives and questioning assumptions in order to arrive at the most accurate conclusions. By consistently asking questions and seeking to understand, rather than to simply validate one's own beliefs, researchers can develop a healthy skepticism that compels them to examine various viewpoints in order to form a comprehensive understanding of their subject matter. By approaching research problems with a curious mindset, we can constantly challenge our own thinking, reducing the risk of adopting a myopic perspective.

Another technique to ensure objectivity is to actively seek out disconfirming evidence. This counterintuitive approach requires researchers to identify information that contradicts their current beliefs or hypotheses. In doing so, they force themselves to consider alternative perspectives and explore contradictory points of view. This process can provide a more balanced perspective on the issue at hand, revealing nuances and complexities that might have otherwise been overlooked. To make this technique more effective, researchers can actively solicit critical feedback from colleagues or engage in structured debates in which each side must argue for the opposing viewpoint. This exercise can hone the ability to think objectively and consider multiple angles.

However, for objectivity to be truly cultivated, it must manifest not only in thought, but in action as well. Actively pursuing diverse sources of information and engaging with a wide range of perspectives on one's area of research can provide greater insight and understanding. This might require collaborating with experts from contrasting disciplines or actively seeking out publications and sources that challenge one's own perspectives. By intentionally exposing oneself to "uncomfortable" ideas or evidence that contradicts our personal beliefs, we develop a more well-rounded understanding of our subject and reinforce the importance of objectivity in our research practices.

In addition to these techniques, simply being aware of potential cognitive biases and exercising mindfulness can also aid in the pursuit of objectivity. Taking the time to reflect on one's thought process, re-evaluating decisions, and engaging in a continuous dialogue with one's own beliefs can help identify potential areas of bias in our research. Practices such as meditation, journaling, and developing metacognitive skills can assist in fostering a heightened level of self - awareness, ultimately enabling researchers to recognize and mitigate the impact of cognitive bias on their work.

Cultivating objectivity demands dedication and effort, as it requires researchers to continuously question their beliefs and assumptions. However, as a singular beam of light is refracted through a prism, revealing the complex spectrum of color that lies within, so too can we uncover the intricate layers of knowledge that exist when we examine our ideas through the lens of objectivity.

By embracing these techniques, such as employing curiosity, seeking out disconfirming evidence, exposing ourselves to diverse perspectives, and practicing mindfulness, we can begin to shed the trappings of personal bias and uncover the deeper connections and insights inherent in the world around us. This dedication to objectivity elevated the work of Darwin, Hopper, and countless other scientists, and it continues to drive pioneering research to this day. In doing so, we not only unlock the vibrant spectrum of understanding lying beneath the surface, but also contribute to the ever - evolving tapestry of human knowledge and innovation.

The Beginner's Mind: Benefits and Methods of Approaching Research with Fresh Eyes

One of the most profound yet overlooked aspects of achieving groundbreaking insights in research is embracing the concept of "the beginner's mind," also known as Shoshin in Japanese Zen Buddhism. This term refers to the attitude of approaching a topic without preconceived ideas, letting go of the expert's mind, and being open to new experiences. The beginner's mind is characterized by curiosity, humility, and receptivity, which enables the researcher to see possibilities that might be otherwise obscured by entrenched beliefs and assumptions.

The history of science is replete with instances where adopting a beginner's mind made the difference between stagnation and innovation. For instance, Sir Isaac Newton, who ushered in a new era of physics and mathematics, wrote: "I seem to have been only like a boy playing on the seashore, and diverting myself in now and then finding a smoother pebble or prettier shell than ordinary, whilst the great ocean of truth lay undiscovered before me." This statement reflects his worldview of not only recognizing the vastness of the unknown but also the willingness to explore it with curiosity and humility.

Another striking case is that of Albert Einstein, who was known for his insatiable curiosity and his ability to approach complex problems with a sense of wonder. Famously, Einstein came up with his groundbreaking theory of relativity by engaging in what he called "thought experiments," in which he imagined riding a light beam and visualizing the implications of that perspective. This imaginative and playful approach enabled him to challenge established beliefs and push the boundaries of scientific knowledge.

To harness the power of the beginner's mind, it is crucial to develop specific strategies and techniques that enable researchers to shed their preconceived ideas and nurture an open, receptive attitude. Some of these methods include:

1. Questioning assumptions: Actively challenging the default assumptions and seeking out alternative viewpoints can reveal new aspects of a problem that were previously hidden. For example, by questioning the belief in absolute time and space, Einstein delved deeper into the fabric of the universe and unveiled a radically new understanding of reality.

2. Cultivating curiosity: By fostering a sense of wonder and fascination with the unknown, researchers can embrace the joy of exploration and become portals for new insights. Techniques such as journaling, brainstorming, and mind - mapping can encourage free - flowing thought and generate new avenues for investigation.

3. Seeking out diverse experiences: By engaging with different fields of study, interacting with people from diverse backgrounds, and exposing oneself to various cultures and beliefs, researchers can develop a richer, more imaginative mental landscape. This broader perspective can propel them toward novel associations and unexpected areas of convergence.

4. Embracing vulnerability: Recognizing the limits of one's own knowledge and the inherent uncertainty in any exploration can free researchers from the shackles of hubris and encourage them to take risks, learn from failures, and stay open to new data and interpretations.

In deploying these methods, researchers can not only broaden their intellectual horizons but also forge more meaningful connections with their peers and mentors. By embracing the spirit of the beginner's mind in their collaborations, they can foster a culture of shared curiosity and mutual learning that propels them toward a collective vision of breakthrough discoveries.

By approaching research with fresh eyes, one discovers that the staggering complexities of the world are but an invitation to an endless dance of inquiry, imagination, and revelation. Like the Zen master who sees wonders in every falling leaf and the child who marvels at the heavens, the researcher who adopts the beginner's mind transcends the confinements of habitual thought and glimpses the radiant tapestries of interwoven truth that unite the myriad fibers of existence. And in those luminous moments of clarity, they become harbingers of a more enlightened, more connected, more awe - inspiring understanding of the universe - an understanding that is only made possible by the boundless curiosity and humility of the beginner's mind.

Case Studies: Examples of Overcoming Cognitive Biases in Famous Research Breakthroughs

Throughout the history of scientific breakthroughs, researchers have had to confront and overcome cognitive biases that may have otherwise hindered progress. The following case studies illustrate how famous individuals in the scientific community were able to recognize and challenge their biases to achieve groundbreaking results.

Case Study 1: Watson and Crick's DNA Discovery

James Watson and Francis Crick are known for their revolutionary discovery of the double helix structure of DNA, the molecule that carries genetic information. This breakthrough would not have been possible if the two scientists hadn't overcome their cognitive biases, particularly confirmation bias.

Initially, Watson and Crick were inclined to believe that DNA had a triple helix structure, and they were unconsciously seeking evidence to confirm this belief. By doing so, they ignored contradicting information that pointed to a double helix structure. It wasn't until they began sharing ideas with other colleagues, like Rosalind Franklin, that they were exposed to new perspectives and data that contradicted their preconceived notion.

Once they recognized the impartiality in their approach, Watson and Crick re-evaluated their initial hypothesis, opening up their minds to the possibility of alternative structures and eventually arriving at the now - famous double helix model.

Case Study 2: Charles Darwin and the Origin of Species

Charles Darwin, the father of the theory of evolution, did not arrive at his groundbreaking ideas without overcoming several cognitive biases. One crucial bias was the availability heuristic, which led him to perceive data that supported his beliefs as more important or prevalent than they might have been in reality.

In crafting his theory, Darwin examined the fossil record and observed a vast number of extinct species. This led him to the belief that species were not fixed entities, contrary to the prevailing notion of the time, but would change and adapt over time. The fossil record provided Darwin with a rich, but incomplete, record of the history of life on Earth.

To challenge the availability heuristic, Darwin spent decades gathering data from diverse fields such as geology, embryology, comparative anatomy, and biogeography. In doing so, he was able to obtain a more comprehensive understanding of the natural world and substantiate his theory.

Case Study 3: Albert Einstein and General Relativity

When Albert Einstein first proposed the idea of general relativity, it contradicted the established understanding of physics at the time. The dominant paradigm, based on Isaac Newton's laws of motion, posited that time and space were absolute entities, independent from one another.

Before Einstein could develop his revolutionary theory of relativity, he had to overcome the anchoring bias, the tendency to rely too heavily on the first piece of information encountered within a given context. In this case, the context for Einstein was the body of existing knowledge about gravity and motion.

By stepping back from the established Newtonian worldview, Einstein was able to explore the relationship between space and time more deeply. He realized that the two were intertwined, with matter causing spacetime to curve and resulting in the phenomenon we experience as gravity. This bold shift in thinking would not have been possible had Einstein not been conscious of and overcome his cognitive biases.

In each of these cases, the great minds of our past have managed to push the boundaries of human understanding by confronting and overcoming their cognitive biases. Watson and Crick, Darwin, and Einstein all demonstrated the power of cognitive resilience, as each individual entertained contradictory or novel perspectives that, in turn, allowed them to make revolutionary discoveries.

These stories remind us to stay vigilant in recognizing and challenging our cognitive biases as we strive for scientific advancement. Embracing a beginner's mind, seeking diverse perspectives, and engaging in self-reflection can all support our quest for unbiased and objective inquiry, fostering an environment conducive to breakthroughs in our understanding of the world around us.

The Role of Feedback and External Input in Overcoming Biases and Enhancing Objectivity

Scientific research is characterized by its pursuit of objective knowledge. Yet, even the most rigorous minds are susceptible to cognitive biases that threaten the validity of their discoveries. With increasing recognition of this reality, researchers are turning towards a potent method for mitigating biases and enhancing objectivity: external feedback and input.

A well-known example of the importance of external feedback comes from the story of Albert Einstein and his general theory of relativity. As brilliant as Einstein was, the physicist Erwin Freundlich corrected a vital calculation mistake he initially made while deriving his revolutionary theory. The partnership between the two ultimately led to the successful prediction of the gravitational lensing effect, a cornerstone of modern astrophysics.

Another case study in the realm of molecular biology exhibits the need for diverse perspectives in research. The discovery of the DNA double helix by Francis Crick and James Watson relied heavily on the X-ray diffraction data provided by Rosalind Franklin and Maurice Wilkins. These four scientists approached the problem with complementary techniques which, when combined, enabled the Nobel Prize-winning revelation that changed the course of genetics research.

Allowing external feedback from diverse sources can not only iron out errors but also highlight assumptions that researchers might not even realize they are making. Charles Darwin's groundbreaking work, "On the Origin of Species," was meticulously researched and refined over two decades, during which he sought input from numerous peers and experts. This feedback helped Darwin identify assumptions he had failed to question, strengthening the eventual publication and securing its status as a cornerstone in modern evolutionary theory.

Researchers must deliberately establish channels for obtaining external input, as our own cognitive biases often cloud our judgment and prevent the identification of our hidden assumptions. Indeed, actively seeking feedback from diverse perspectives is a crucial strategy for overcoming biases. This can be achieved through various methods, including peer review, interdisciplinary collaboration, and conducting workshops and seminars.

A case in point is the development of the internet. One might be surprised to learn that the internet's backbone hinges on a collection of humble documents known as Requests for Comments (RFCs). These RFCs, authored by researchers and engineers working on various aspects of the burgeoning technology, were circulated for input from the wider community. This collaborative culture of seeking feedback and input allowed for rapid iterations and refinements, which catalyzed the growth of the global network we know today.

Despite the value of external input, researchers must balance this exchange with the nurturing of independent creativity. This delicate balance is exemplified by the story of Jonas Salk, the developer of the first effective polio vaccine. While Salk frequently engaged with the scientific community, he maintained a degree of secrecy during the vaccine's development to avoid compromising his innovative approach. In doing so, he successfully navigated between an openness to critique and the need for autonomous creative space.

Ultimately, incorporating feedback and external input in the research process requires an attitude of humility and a willingness to learn. By embracing the insights and expertise of others, we may transcend the confines of our subjective points of view and forge a bridge to objectivity.

As we progress through this exploration of visionary minds and techniques, it is essential not to lose sight of the broader context. The virtues of overcoming cognitive biases and integrating external input are best employed within interdisciplinary, collaborative research endeavors that unite disparate fields of knowledge. In tandem, these powerful strategies for intellectual inquiry foster the ideal conditions for insights that illuminate the path towards innovation and discovery.

Mindfulness and Self-awareness: Building Mental Habits to Combat Biases

Mindfulness and self-awareness are powerful mental habits that can be utilized to combat the cognitive biases that often inhibit objective research. Mindfulness is a mental state in which one intentionally focuses their attention on the present moment, while also maintaining a non-judgmental and open-minded awareness of their thoughts, feelings, bodily sensations, and surrounding environment. Self-awareness, on the other hand, refers to the conscious knowledge of one's own character, feelings, motivations, and desires. Together, these concepts can be harnessed to overcome confirmation bias, recency bias, and other obstacles to objective, fearless thinking.

Consider the power of mindfulness in the context of a major historical discovery: the realization that the Earth revolves around the Sun. Nicolaus Copernicus, a mathematician and astronomer, was able to contradict an entrenched system of belief (geocentrism) only by remaining grounded in the present moment and avoiding the seductive lure of consensus thinking. By cultivating mindfulness, he was able to carefully observe and analyze astronomical data with fresh eyes, free from the biases that could have clouded his judgment.

Self-awareness, too, has played a critical role in some of the great discoveries of human history. Charles Darwin, for example, was highly self-aware of his own expectations. Rather than allowing these expectations to shape his interpretation of the data collected during his voyage on the HMS Beagle, Darwin actively sought out disconfirming evidence that challenged his preconceptions. This introspective diligence led to his groundbreaking theory of evolution by natural selection.

One of the most effective ways of building mindfulness and self-awareness is through regular meditation practice. Researchers from the University of California, Santa Barbara, demonstrated that even just two weeks of mindfulness meditation training led participants to score significantly higher on a cognitive test that required them to ignore distracting information and focus on the task at hand. As we can surmise, cultivating such a discipline would be invaluable in the research process, allowing practitioners to maintain objectivity and mental clarity in the face of potential distractions.

Another technique to enhance self-awareness involves maintaining a personal reflection journal. By consistently recording thoughts, emotions, and impressions associated with research findings, individuals can develop a deeper understanding of their internal narrative and thought processes. In doing so, they can recognize instances where cognitive biases may be influencing their decisions and subsequently implement corrective measures accordingly.

A memorable example of this strategy is Albert Einstein's famed thought experiments. By engaging in imaginative exercises such as envisioning himself riding alongside a beam of light, Einstein was able to mentally interrogate his own assumptions and employ unconventional perspectives, thus paving the way for his revolutionary theory of relativity.

As with meditation, numerous studies have demonstrated the efficacy of journaling in promoting self-awareness and mitigating cognitive biases. For instance, research conducted at the University of Utah suggested that individuals who participated in an expressive writing intervention exhibited a reduced propensity for the sunk cost fallacy, a common cognitive bias that often leads individuals to persist with unproductive decisions due to previously invested resources. Thus, dedicating 10-20 minutes per day to personal reflection can lead to lasting improvements in identifying and tempering cognitive biases.

To unlock the full potential of mindfulness and self-awareness, it is essential to combine these practices with a genuine commitment to intellectual curiosity and open-mindedness. As history has shown us, the most transformative discoveries and innovations often arise when individuals transcend their preconceived ideas in favor of flexibility, adaptability, and imagination.

In short, embracing mindfulness and self-awareness can enable researchers to approach problems with a refined sense of mental clarity, focus, and objectivity. By consistently engaging in introspective exercises such as meditation and journaling, researchers can hone the mental habits necessary to identify and mitigate cognitive biases. This paves the way for breakthrough insights, which, more often than not, require adopting perspectives beyond the confines of conventional thinking. By undertaking this journey toward objectivity, researchers can become more equipped to leverage their diverse interdisciplinary knowledge in pursuit of cutting-edge discoveries which, like Copernicus or Einstein's groundbreaking insights, could reframe our understanding of the world and our place within it.

Conclusion: Embracing the Journey toward Objectivity and Beginner's Mind in Research

The quest for objectivity begins by adopting a beginner's mind - an open, unbiased perspective that acknowledges the limits of our knowledge while encouraging exploration and reinterpretation. This mindset is epitomized by scientific luminaries such as Einstein, who revolutionized our understanding of the universe through his relentless pursuit of foundational truths. Adopting a radically open perspective is not only an essential step towards conducting robust, data-driven research but can also lead to groundbreaking ideas that challenge the very fabric of our understanding.

A key strategy to maintain objectivity lies in identifying and neutralizing cognitive biases that subtly creep into our decision - making processes and hinder our ability to perceive reality as it is. Confirmation bias, for example, might impel us to subconsciously seek out supportive evidence while neglecting the larger context. We must actively cultivate self-awareness and mindfulness, intentionally scrutinizing our assumptions and inclinations. Embracing the input of fellow researchers and integrating constructive feedback can help unveil the blind spots that elude our biased vision.

It is also important to recognize that objectivity is not a solitary endeavor. The treasure troves of insights gleaned from the intersection of diverse perspectives underscore the importance of challenging one's worldview with alternative viewpoints. As we learn from the story of Watson and Crick, their dogged determination and receptiveness to novel ideas culminated in the breakthrough discovery of the double - helix structure of DNA. This exemplifies the power of embracing intellectual curiosity and collaborative inquiry in the pursuit of objectivity.

As we proceed on our journey towards unbiased and robust research, we must not become daunted by the sheer enormity of the undertaking. Grand strides in our understanding of the cosmos have always stemmed from our collective ability to challenge and rethink our long-held convictions. We must learn to embrace adversity as an opportunity for growth, build resilience, and cultivate rituals that inspire creativity and perseverance.

The landscape of objective research is ever - evolving. A kaleidoscope of new developments and perspectives constantly emerges, molding the scientific frontier and inspiring the trailblazers of tomorrow. As we grapple with the inherent complexities and intricacies of our universe, we must continue to foster the spirit of the beginner's mind - a mind that is open, curious, and unyielding in its pursuit of truth.

Chapter 6

Causal Efficacy: Mastering First Principles Reasoning for Robust Discoveries

As researchers, we often find ourselves trying to decipher elaborate scientific mysteries. It can be daunting to recognize that the nuances of the natural world elude us. But, what if the answer for untangling these mysteries lies not in accepting complexity, but rather, in embracing simplicity? Causal efficacy, or first principles reasoning, is a way of thinking that enables us to harness the power of simplicity to make groundbreaking discoveries.

At its core, the first principles approach involves deconstructing complex concepts into their basic building blocks and then reassembling these elements to form novel ideas. This ability to reason from rudimentary principles allowed the likes of Galileo Galilei, Johannes Kepler, and Sir Isaac Newton to challenge longstanding beliefs and achieve intellectual triumphs that would reshape our understanding of the physical world.

One of the defining moments in the history of physics can be attributed to Galileo's famous Leaning Tower of Pisa experiment. Contrary to Aristotle's claim that heavier objects fall faster than lighter ones, Galileo hypothesized that all objects fall at the same rate. To prove his theory, he carefully broke down the problem into its constituent elements, most notably the forces acting upon an object in freefall. Through a combination of experimentation and logical reasoning, Galileo's first principles approach led him to the discovery that objects fall at the same rate regardless of their mass, shattering a 1500 - year - old misconception.

Kepler, too, provides us with a sterling example of why first principles reasoning often leads to extraordinary scientific revolutions. Instead of taking the prevailing view that planetary orbits had to be circular, Kepler stripped away his received notions about the nature of the cosmos and started afresh. By analyzing Tycho Brahe's meticulous observations, Kepler recognized that the motion of planets followed elliptical, rather than circular, paths around the Sun. This groundbreaking insight overturned the traditional Ptolemaic model of the universe and laid the groundwork for Newton's laws of motion and universal gravitation.

On the topic of Newton, one need not look any further than his classic apple revelation to understand the value of first principles thinking. Rather than accepting the commonly held presumption about gravity's sudden cutoff - which asserted the force of gravity was nonexistent beyond Earth's immediate vicinity - Newton reframed the problem from the ground up. By constructing a new mental model for gravity that depicted gravitational force as a continuous phenomenon, Newton's reasoning directly challenged the prevailing conception of gravity and ultimately led to one of the most transformative developments in the history of science: the universal law of gravitation.

These pioneers of science and many others demonstrate that first principles reasoning can be a powerful tool in enabling us to push the boundaries of what is known. But how, you may ask, can we put these insights into practice in our own research endeavors?

One key strategy is to become skilled in breaking down complex problems into their constituent parts. By taking a problem apart and laying out its fundamental elements, we can identify the core dynamics at play, distinguish between assumptions and evidence, and ultimately reconstruct the problem in a way that illuminates innovative solutions.

Another crucial aspect of first principles thinking is the ability to identify and challenge underlying assumptions. Recognizing that certain assumptions may constrain our thinking or lead us astray enables us to adopt a more flexible mindset and question seemingly incontrovertible beliefs.

Lastly, using analogies and other logic-based structural analysis methods can aid us in reconstructing problems and finding new connections between different concepts. By drawing parallels between disparate phenomena or applying an entirely new framework to our analysis, we can craft fresh insights and bring about the next big breakthrough.

As we march forward into an ever - complexifying world of scientific inquiry, the gift of first principles thinking becomes even more critical to our success. Embracing the art of causal efficacy can open doors to new horizons that were previously beyond our reach. Raise your sights to the lofty heights of Galileo, Kepler, and Newton and take heart in the knowledge that ingenuity - fueled by the enduring power of simple, foundational reasoning - knows no bounds. The universe, in all its baffling complexity, awaits those diligent enough to mine for the golden nuggets of truth hidden beneath the surface.

In the journey of unlocking scientific mysteries, armed with the wisdom of history's most accomplished visionaries, we must acknowledge that tomorrow's groundbreaking ideas can only be born through the synthesis of various disciplines. Indeed, as Sir Francis Bacon once said, "The world is but a perennial movement from the heterogeneous to the homogeneous, and back again from the homogeneous to the heterogeneous."

Introduction to First Principles Reasoning: Origins and Importance in Scientific Discoveries

In the realm of scientific discovery, an often overlooked yet immensely powerful tool employed by history's greatest thinkers is the process known as "First Principles Reasoning." Meticulously peeling away layers of beliefs, assumptions, and convention, this technique drives the mind to the bedrock of reality, exposing the essential, irreducible axioms upon which our universe is built. From there, through careful analysis, the scaffolding of new concepts, hypothesis, and ultimately, world - changing discoveries can be erected. As we embark upon this exploration of First Principles Reasoning, we will traverse the historical landscape of groundbreaking innovations, gleaning valuable insights that can be applied to strengthen our own intellectual pursuit in the noble domain of scientific research.

The journey begins with a discussion of origins, for it is in the footprints of history's titans that we find the glowing embers of wisdom we seek to ignite. While the seeds of First Principles Reasoning can be traced back to the days of early Greek philosophers such as Aristotle and Socrates, its true application to the field of science did not fully manifest until the modern era. It was during the age of Enlightenment, with the emergence of scientific giants such as Johannes Kepler, Galileo Galilei, and Sir Isaac Newton, that this cognitive tool became crystallized as a driving force within the hallowed halls of intellectual endeavor.

The importance of First Principles Reasoning cannot be understated. For Kepler, it was through relentless examination of the natural world and a willingness to challenge entrenched assumptions about celestial motion that he led him to formulate his laws of planetary motion - a direct challenge to the millennia - old belief in geocentrism. For Galileo, careful observation of the moons of Jupiter also led to a staunch conviction in the heliocentric model of the solar system, despite the severe personal and political consequences he faced for voicing his opposition to the dominant views of his time.

And finally, perhaps the most famous and profound example of First Principles Reasoning in scientific history can be found in the life and work of Sir Isaac Newton. His revolutionary work not only synthesized the genius of his predecessors - such as the laws of motion and universal gravitation - but also tore down the walls of understanding that had existed for centuries, laying the foundation upon which classical physics would be built. By starting with a simple question - "What keeps the Moon in orbit around the Earth?" - Newton meticulously dismantled the gauntlet of conventional wisdom, unveiling a breathtakingly elegant and comprehensive view of the cosmos that had never before been glimpsed by human eyes.

These luminaries of scientific thought represent but a few shining examples of what can be achieved through First Principles Reasoning. The elegance of this approach lies in its insistence on simplicity, striving to reduce complex concepts down to a core set of fundamental principles from which new ideas and insights can emerge. By embracing this modality, we are not only paying homage to the intellectual giants who came before us but are also rising to join their esteemed ranks, becoming active participants in this grand tapestry of discovery upon which our world relies.

As we continue our exploration into the intricacies of First Principles Reasoning, we encourage you, the reader, to view this ancient yet timeless tool through the lens of a detective seeking clues, a philosopher seeking truth, and a scientist seeking understanding. Through careful examination of historical examples, identification and challenging of assumptions, and the skillful wielding of this formidable intellectual weapon, we join together in the pursuit of that most elusive and tantalizing of treasures - the next great scientific breakthrough.

And who knows, in the quiet moments of deep reflection, when our thoughts emerge unbidden and our minds venture forth into untamed realms, perhaps we too can catch a glimpse of the divine spark that set the planets in motion or the ethereal breath that whispered the secrets of gravity to the wind. History awaits the next visionary willing to cast aside convention, and embrace the infinite possibilities that lay hidden within the realm of First Principles Reasoning. Will that visionary be you?

Breaking Down Complex Problems: Deconstructing Concepts into Fundamental Elements

One of the most remarkable examples of breaking down complex problems comes from the field of physics, particularly the study of gravity. Issac Newton was faced with the seemingly impenetrable mystery of how celestial objects maintained their orbits. The triumphant outcome of his collaborative campaign to decode the problem culminated in the development of the three laws of motion and the law of universal gravitation.

Newton's approach, grounded in first principles reasoning, involved deconstructing the concept of gravity by isolating the fundamental elements of the problem, such as mass, force, and acceleration. By breaking down these elements, he managed to derive relationships between them and identify the principles governing their behavior. This, in turn, led to a better understanding of planetary motion and paved the way for modern physics.

Similarly, the discovery of the structure of DNA exemplifies the effective dissection of complex problems. James Watson and Francis Crick, the scientists credited with the discovery, faced a convoluted puzzle of nucleotide sequences and chemical bonds. It was only through iterative experimentation and re - examination of the underlying components, such as the famous base - pairing rule, that they eventually discerned the elegant double helix structure.

A practical approach to breaking down complex problems involves the following steps:

1. Define the problem: For a clear understanding of the issue at hand, it is imperative to formally state the problem. Doing so elucidates the components that need further investigation.

2. Identify the underlying elements: With an explicit problem definition, it becomes easier to identify the fundamental elements that constitute the problem.

3. Analyze the relationships between the elements: Once the elements of the problem are isolated, investigate the relationships and interactions between them. This step requires a deep understanding of each element's behavior and its role in the broader problem.

4. Synthesize a new framework: Following a thorough analysis of the elemental relationships, reconstruct a coherent framework that adequately addresses the complexities of the problem. This reassembled structure will likely reveal new insights and possibilities.

Throughout history, groundbreaking thinkers have employed the method of deconstruction to drive their great discoveries. Their tenacity to dissect the intricate fabrics of the universe and lay bare its most fundamental elements has gifted humanity with substantial leaps forward in understanding and knowledge.

Aristotle, in his quest to understand the nature of reality, posited the concept of substance - the fundamental, unchanging nature of a thing - which led to significant advancements in both science and philosophy. Charles Darwin, through rigorous observation and comparison of different species, unveiled the revolutionary concept of natural selection by breaking down the process of evolution into its core components. Marie Curie, while hunting for the source of radioactivity, tirelessly sought the most basic particle causing the phenomenon, ultimately unveiling the presence of polonium and radium.

In conclusion, the method of breaking down complex problems serves as a potent tool to illuminate hidden facets of knowledge and propel scientific discovery to new heights. As we forge ahead in our pursuit of uncharted territories, it becomes increasingly important for researchers to view intricate concepts through the lens of deconstruction to proactively confront the challenges and limitations of hypothesis generation. Only then can we collectively unravel the profound and mysterious tapestry of the universe that so innately captivates the curious human spirit.

Identifying and Challenging Assumptions: The Key to Overcoming Limiting Beliefs

Assumptions are deeply embedded in our thinking process and are vital for reasoning and decision - making. We rely on them to navigate the complexities of the world, to fill gaps in our knowledge, and to simplify information. By doing so, however, we sometimes undermine our ability to achieve breakthrough thinking in our research endeavors, as unchallenged assumptions can hinder the discovery process and maintain the status quo. Identifying and challenging these assumptions is essential, because it can lead us to question and reevaluate the fundamental principles on which our research is based. Only by breaking free from limiting beliefs can we develop new understandings and contribute to advancing the frontiers of human knowledge.

The scientific method itself is built upon questioning assumptions. As researchers, we are constantly searching for empirical evidence to support our hypotheses, measure the unknown, and disprove the false. This process safeguards us from the possible pitfalls posed by cognitive biases and errors, while encouraging us to adapt our ideas in the light of newly discovered facts. However, the act of questioning assumptions does not necessarily come naturally, and must be intentionally nurtured and cultivated.

Take the story of Ignaz Semmelweis, for example. In the mid - 19th century, Semmelweis, an Austrian doctor, observed that women who gave birth in a hospital had a significantly higher mortality rate than those who gave birth at home. His investigation revealed that doctors were often going directly from handling corpses in the morgue (to study cadavers and perform autopsies) to delivery wards without washing their hands in between. This contradicted the conventional medical wisdom of the time, which believed that bad air, or miasma, caused disease transmission.

By challenging the prevailing assumptions, Semmelweis proposed that some kind of "cadaverous matter" on doctor's hands was responsible for the deadly cases and puerperal fever. He conducted rigorous experiments to test his hypothesis, instigating a strict handwashing regimen for clinicians. Remarkably, mortality rates plummeted, and Semmelweis' discovery laid the groundwork for germ theory and antiseptic practices.

To question and challenge assumptions, as Semmelweis did, we must first be able to identify them. This can be facilitated through several techniques: inquiry, analysis, and reasoning. Inquiry constitutes asking why, how, and when - forcing us to confront the foundations of our beliefs. Analysis entails meticulously breaking down a problem into its constituent elements and isolating each assumption to assess their individual validity. Reasoning provides the refined logical framework upon which to differentiate between reasonable and unreasonable assumptions.

Once assumptions are identified, challenging them becomes fact - driven, evidence - based, and outcome - oriented. The act of questioning an assumption necessitates providing alternative perspectives and potential counterarguments. By engaging in this process, we may find that our initial assumption was, in fact, accurate and reliable. In other cases, we may unearth fresh insights or generate novel connections, which push our research in unforeseen directions, allowing for breakthrough discoveries to emerge.

The process of challenging assumptions is less about questioning everything for the sake of it than it is fostering an open - minded, adaptable, and inquisitive environment conducive to the pursuit of new knowledge. The Socratic method, for instance, encourages dialectical exchange and the use of logic and reasoning, often overturning traditional assumptions to unearth deeper truths. This ancient technique remains valuable in the modern age, as it cultivates positive intellectual dissent, which can foster genuine innovation and understanding.

In summary, by adopting Semmelweis' spirit of challenging assumptions, we can surpass limiting beliefs, foster creative thinking, and lead to groundbreaking research. The simple acts of identifying and questioning our assumptions empower us to overcome cognitive obstacles, allowing us to navigate the unknown territory of discovery with intellectual courage. By carrying this intrepid mindset into the future, we stand ready to embrace novel ideas and methodologies as they emerge and weave them into the rich tapestry of human understanding.

Using Analogies and Logic Based Structural Analysis: Reconstructing Problems for Novel Insights

In order to uncover novel insights and solve complex problems, researchers have long relied on the power of analogies and structural analysis. By examining their own fields through the lenses of other disciplines or seemingly unrelated subjects, they are able to discover new perspectives and find innovative solutions. Indeed, many groundbreaking discoveries have resulted from the ability to draw striking parallels between seemingly unrelated phenomena and concepts.

Consider, for example, the work of the eminent physicist Richard Feynman. Feynman's curiosity led him to explore a wide variety of subjects, and his penchant for analogy was central to his innovative thinking. In his research on atomic particles, Feynman was inspired by a shopkeeper's technique for counting items. As a result, he developed the groundbreaking Feynman diagrams, which revolutionized the way scientists understand and model subatomic particle interactions.

This fascination with analogy is not limited to the realm of physics. Biologists, too, have made significant strides by embracing analogous thinking. One prime example is the case of DNA. When James Watson and Francis Crick were struggling to understand the structure of DNA, they drew upon their knowledge of architecture and chemistry, specifically analogies with the structure of complex organic molecules. This interdisciplinary approach ultimately allowed them to "see" the double helix structure of DNA, which is now the backbone of genetics and molecular biology.

So, how can researchers and scientists apply these techniques of using analogies and logic-based structural analysis to reconstruct problems and unlock novel insights in their own work? One method is to expose oneself to a range of different disciplines and fields, as doing so can provide a broader foundation of knowledge from which to draw parallels. By examining the structure and relationships between elements in other domains, researchers may uncover similarities that shed light on their own subject matter.

Another strategy is to consider problems from multiple angles, actively looking for analogies and patterns between seemingly unrelated topics. This can involve incorporating elements such as visualization, mind mapping, or sketching, in order to create a visual representation of the problem and its components. By seeing a problem in a new light or from a different perspective, researchers can often identify novel insights or creative solutions that might elude a more linear, rigid analysis.

Furthermore, researchers can embrace the practice of "translating" complex concepts from their own field into simpler, more universally understood language. By doing so, they can often uncover hidden connections and parallels to other disciplines, which might ultimately yield innovative discoveries, as was the case for Watson and Crick when they "translated" complex organic chemistry into the more accessible language of architecture.

In order to effectively apply these strategies, it is essential for researchers to cultivate a mindset of curiosity, openness, and receptivity to new ideas and perspectives. By actively seeking out different viewpoints and embracing the possibility that innovative solutions may lie outside of one's own field or area of expertise, researchers can overcome cognitive biases that might otherwise restrict their thinking and stymie their progress.

As researchers begin to unlock these new insights through analogy and structural analysis, it is crucial that they share their findings with others and engage in collaborative learning. Just as James Watson and Francis Crick teamed up to crack the DNA code, modern researchers can forge interdisciplinary partnerships and work together to solve the challenging problems of our time. Such collaboration not only serves to expand the collective knowledge of humanity but also opens the door to groundbreaking discoveries that might otherwise remain hidden in the dark corners of isolated fields.

First Principles Reasoning in Action: Case Studies of Kepler, Galileo, and Newton

The power of first principles reasoning is best observed through the actions of history's great minds. When examining the lives and works of renowned scientists such as Johannes Kepler, Galileo Galilei, and Isaac Newton, we witness the transformative nature of questioning the status quo and distilling complex problems into their most basic, elemental truths. Each of these visionaries drew upon the essence of first principles reasoning, challenging assumptions and constructing novel frameworks of understanding that ultimately revolutionized their respective fields.

Johannes Kepler, a true trailblazer in the field of astronomy, was relentless in his pursuit of the understanding of planetary motion. Mineral dust still clung to centuries of unquestioned Ptolemaic assumptions when Kepler pursued the idea that there was an unseen simplicity in character to the orbits of the planets. Emboldened by the accurate observations of his late mentor, Tycho Brahe, Kepler peeled away the layers and discarded old complexities. He hypothesized that planets' orbital paths could be predicted by geometric shapes, a conclusion antithetical to the longstanding belief in perfectly circular orbits. His tenacious analysis and mathematical rigor, operating from basic concepts of the planets' speed and the heliocentric model, undeniably led him to discover the three laws of planetary motion, which still underpin modern astrophysics.

Galileo Galilei, another titan in the field of astronomy, as well as physics, similarly employed first principles reasoning in his work. Armed only with the tools of his own intuition and primitive telescopes, Galileo embarked on a journey to expose the fallibility of age-old conceptions about our universe. Defying popular belief, he systematically dismantled the Ptolemaic, Earth-centered model by observing the moons of Jupiter, proving they were not in orbit around Earth. Further, Galileo applied first principles reasoning to the study of motion, abandoning Aristotle's conclusion that heavier objects fall faster than lighter ones. Instead, Galileo reduced the issue to its core elements, considering vital aspects such as mass and air resistance, and developed the universal law of free fall. It was through his unshakable demonstrations, grounded in basic truths, that allowed him to construct a more accurate, comprehensive understanding of the natural world.

Isaac Newton, the father of classical mechanics, revolutionized the scientific community's perception of the very forces governing our universe. His rigorous application of first principles reasoning and methodology exemplifies these core philosophies, most notably in his derivation of the laws of motion and universal gravitation. Newton held firmly to the belief that universal laws of nature were expressed in mathematical terms and pursued these immutable first principles with unwavering conviction. By dissecting the puzzle of planetary motion into its constituent, calculable elements, Newton unlocked the celestial-sanctioned secrets that Kepler first glimpsed and Galileo steadfastly pursued. Building upon the works of these intellectual giants, and rooted in simplicity and precision, Newton wove together the three laws of motion and discovered the law of universal gravitation, crucial elements that led to the development of his magnum opus, the "Principia."

Although each of these luminaries worked within their respective domains, their lives and achievements share a cohesive narrative: the application of first principles reasoning as a potent tool for discovery, enabling them to transcend the bounds of established knowledge and birth revolutionary insights. The examples of Kepler, Galileo, and Newton remind us of the power of stripping away complexity and returning to foundational truths as the cornerstone of scientific progress. It is in the humble interrogation of assumptions and the synthesis of logic and creativity that we unlock the potential for astounding breakthroughs and visions into realms previously unexplored.

And so, as we continue to forge our paths and confront the challenges faced by modern research, we must remember the lessons imparted by these trailblazers and carry with us the spirit of first principles reasoning. For it is in this inquisitive mindset, rooted in logic and dedication to elemental truth, that we will find the key to unlock the secrets of our universe and drive the engines of innovation far beyond the horizons we imagine today.

Integrating First Principles Reasoning into Research Methods: Tools and Techniques for Robust Discoveries

Integrating First Principles Reasoning into research methods is not only a way to reinvigorate traditional scientific inquiry; it also provides a robust framework for discovering new insights and fostering innovation. The process begins by breaking down complex problems into their most fundamental elements, identifying and challenging critical assumptions, and using analogies and logic-based structural analysis to reconstruct these problems in a new light. These techniques can be applied across various disciplines, helping researchers cultivate a more critical and creative mindset that cuts through the noise of conventional wisdom.

One essential step in integrating First Principles Reasoning into research is to identify the most basic and necessary components of a problem. This requires a level of abstraction that goes beyond simple surface analysis and pushes researchers to delineate the boundaries between fundamental principles and superficial assumptions. Take, for example, the development of the atomic model. Early scientists did not simply accept the idea that matter was composed of various elements; instead, they sought to uncover the basic principles governing the seemingly infinite variety of substances in the world. Through a recursive process of hypothesis formulation and experiment design, they identified the existence of atoms and subatomic particles, providing the foundation for modern chemistry and physics.

Another crucial technique for integrating First Principles Reasoning into research is to challenge assumptions. This often requires a willingness to question widely accepted norms that may be more a product of historical momentum than rigorous demonstration. One example of this is the practice of bloodletting in medieval medicine. Many physicians believed that illness resulted from an imbalance of humors in the body, and that removing "excess" blood would restore harmony and promote healing. This assumption persisted for centuries, until researchers began to question the validity of humoral theory and seek more grounded, testable hypotheses about the nature of disease. By challenging the prevailing orthodoxy, they enacted a paradigm shift that paved the way for modern medical research.

The use of analogies and logic - based structural analysis can also be invaluable in clarifying problems in new ways and finding innovative solutions. For instance, the advent of antibiotic resistance created an urgent need for alternative treatments to conventional drugs. Researchers turned to an unlikely source for inspiration: the natural world. By studying the strategies employed by organisms to defend against bacterial infections, they identified novel mechanisms for targeting pathogens, such as the use of bacteriophages (viruses that infect bacteria) and antimicrobial peptides. Drawing analogies between the biological mechanisms at play in these natural defenses and the principles of drug development allowed scientists to synthesize new classes of therapeutics, leading to the birth of a promising field known as "biomimicry."

To successfully integrate these strategies into their research methods, researchers must develop the mental habits that foster the continuous application of First Principles Reasoning. One such habit is to routinely question the validity of assumptions and to probe the limits of their applicability. This can involve practicing "devil's advocate" exercises, in which researchers aim to identify the weak points in their own theories and arguments. Additionally, cultivating curiosity and an interdisciplinary mindset can lead to the discovery of relevant information and insights from fields outside one's area of expertise. Researchers should be encouraged to reflect on their own beliefs and knowledge structures and identify areas where additional learning might be fruitful.

In conclusion, mastering the tools and techniques of First Principles

Reasoning can unlock the potential for groundbreaking discoveries in any field of research. By stripping down complex problems to their foundational elements, challenging assumptions, and utilizing analogies and logical analysis to reframe these issues in innovative ways, researchers can transcend disciplinary boundaries and conventional wisdom. Moreover, by fostering mental habits that embrace curiosity, reflection, interdisciplinary thinking, and intellectual rigor, they can continuously apply this powerful approach to their work, propelling them - and their research - towards the outer limits of human understanding. And as our visionaries have shown us time and again, there are no limits where curiosity, creativity, and rigorous inquiry lead.

The Role of First Principles Reasoning in Unlocking Future Breakthroughs and Advancing Human Knowledge

First principles reasoning, a methodology that involves deconstructing complex problems into their most fundamental elements, has been one of humanity's most powerful tools in unlocking breakthroughs and advancing human knowledge. By re-starting from these core elements and formulating innovative solutions, our civilization's great minds have been able to transcend traditional boundaries and redefine what we once thought was impossible.

One prime example of the power of first principles reasoning lies in the scientific revolution of the 17th century. Classical Greek and Medieval Aristotle's physics dominated scientific thinking for centuries and believed that different objects' movement could only be explained by different forces. However, Galileo Galilei challenged these assumptions and carefully analyzed the physics of motion in a groundbreaking way. Combining mathematics and observation, he laid the groundwork for Newton's laws of motion, which emerged as universal laws of nature and entirely shifted humanity's understanding of the physical world.

In more recent times, we have seen the extraordinary impact of first principles reasoning in the development of the atomic model. Our understanding of atomic structure was shaped by many great minds who proposed and refined different hypotheses. At the heart of each contributory theory was an application of first principles thinking. Bohr's idea of quantized electron orbitals, for instance, was a pioneering breakthrough that shattered traditional models of atomic structure and paved the way for the field of quantum mechanics.

As we look towards the future, it is clear that first principles reasoning will continue to be an indispensable tool in overcoming Earth's most pressing issues. To adequately address the challenges posed by climate change, for example, we must apply a first principles mindset to completely reevaluate our energy systems, reduce our carbon footprint, and develop cost - efficient green technologies.

Elon Musk's company, Tesla, exemplifies this approach by rethinking automotive engineering from the ground up. Instead of adhering to conventional industry wisdom about electric vehicles' limitations, Musk used first principles reasoning to challenge preconceived notions and create a game-changing electric car with impressive range, performance, and design - ultimately making electric cars a viable and attractive option for consumers.

Furthermore, in the rapidly - evolving landscape of biotechnology, first principles reasoning will be invaluable in revolutionizing medicine and healthcare. By analyzing the fundamental principles of genetics, cellular functions, and neuroscience, innovative technologies such as CRISPR gene editing and brain - computer interfaces emerge as promising solutions to a wide range of health issues. These breakthroughs have the potential to transform our overall understanding of biology, address previously incurable diseases, and significantly improve our quality of life.

First principles reasoning is not a mere intellectual curiosity but an essential tool for researchers across various fields aspiring to contribute to human progress. By incorporating this methodology into problem - solving, scientists can transcend short - term trends and surface - level observations, focusing on the root causes of complex problems and synthesizing novel, groundbreaking solutions.

As we embark on the next wave of remarkable breakthroughs, the significance of first principles reasoning becomes increasingly apparent. The implementation of this philosophical approach will undoubtedly contribute to new discoveries in a rapidly - changing world, driving us closer to a future of sustainable living, increased longevity, and unimaginable technological advancements.

So, as the proverbial baton passes from the distinguished visionaries of the past to the brilliant minds of the present and future, let us remember the immeasurable value of first principles thinking. Our collective evolution is contingent upon it. Embracing and nurtured within the boundaries of our inquiries, the principles stand as a reminder that our limitations are, in the end, merely the doorways to profound transformation.

Chapter 7

Individual Creativity and Group Collaboration: Balancing the Equation for Success

The development of the theory of relativity by Albert Einstein is often seen as a quintessential example of individual creativity at work, with the solitary genius coming up with groundbreaking ideas that change the course of human understanding. Einstein's brilliance is indisputable, and individual creativity is a crucial driver of innovation. However, the advancement of knowledge requires more than isolated bursts of inspiration. It demands the collective efforts of many minds, merging together different perspectives and insights, and building upon one another's ideas to propel humanity forward. The complex landscape of modern research necessitates a delicate balancing act between individual creativity and group collaboration, an equilibrium that is essential for achieving pioneering breakthroughs.

Individual creativity breeds originality, helps challenge the status quo, and people explore the unknown. Creative individuals, driven by their curiosity and passion for knowledge, can push the boundaries of understanding and challenge existing paradigms. They can resist conformity, adapt to fresh perspectives, and navigate through uncertainty to uncover new truths. Creativity is an essential element of scientific imagination - the ability to envision possibilities and devise ingenious methods to test hypotheses. Thomas Kuhn's famous conception of "paradigm shifts" in the history of science is often underpinned by individual creativity, the capacity to build new explanatory frameworks that supersede and redefine existing ones.

However, the scientific enterprise is not and should not be a solitary pursuit. The plethora of scholarly disciplines, the complicated nature of contemporary problems, and the limitations of human cognition require collective endeavors, where people with diverse skills and expertise come together to build a shared understanding. Collaboration holds the key to approaching complex problems from different angles, to scrutinizing and extending the scope of individual hypotheses, and to refining and validating experimental designs.

Collaboration can harness the collective intelligence of the group, minimizing the biases and blind spots inherent in any individual's thinking. It facilitates the sharing of resources, data, and methodological tools that enable teams to overcome obstacles. Collaboration also drives the cross-fertilization of ideas, creating intellectual synergy that can push the frontiers of knowledge and technology.

History presents countless examples of successful group collaborations that led to groundbreaking discoveries. Crick and Watson's joint efforts in determining the structure of DNA, the collective work of the Manhattan Project on the development of the first atomic bomb, and the recent international collaboration to discover the Higgs boson particle at CERN are testaments to the power of collective intellect.

Balancing individual creativity and group collaboration is essential for research success. It requires a delicate equilibrium where creative minds can flourish, while also fostering a culture that encourages intellectual exchanges and collaboration. Institutions and research leaders must ensure an atmosphere that nurtures and respects both individual autonomy and collective synergy by fostering openness and curiosity, establishing ground rules for constructive criticism and debate, and promoting diversity in perspectives and backgrounds.

One powerful example of this balance comes from the famous Solvay conferences of the early 20th century, a series of gatherings that brought together leading physicists such as Einstein, Bohr, and Schrödinger to discuss and debate the most pressing questions of their time. These meetings were characterized by an intense exchange of ideas - at times as heated disagreements - but also by mutual respect and admiration. The outcome was a series of incredible advances in our understanding of quantum mechanics and particle physics.

The journeys of Grace Hopper and Katherine Johnson, pioneering women who overcame significant barriers and biases to excel as leaders in their fields, offer valuable lessons on balancing the equation for success. As an integral part of collaborative efforts at the time, they equipped researchers with the understanding of modern programming and calculations vital to numerous projects, including the Apollo moon landing mission. Their leadership style, characterized by intellectual curiosity, adaptability, and inclusivity, played an essential role in shaping the teams that pushed the frontiers of computer science, mathematics, and space exploration.

In a world where research is increasingly defined by the converging and overlapping interests of diverse fields, the skillful balance of individual creativity and group collaboration is more important than ever. The scientific endeavor is not a zero - sum game where individual brilliance must be sacrificed for the common good, or where collaboration stalls creative juices. It is an intricate dance, a process of co - creation where the synergy of disparate minds leads to breakthroughs that reshape our understanding of the universe and our place within it. As we stand on the brink of revolutions in artificial intelligence, biotechnology, and space exploration, the wisdom and resilience of Hopper and Johnson's legacy serve as a guiding light, an invaluable reminder of the principles that propel human ingenuity to ever - greater heights.

The Importance of Balancing Individual Creativity and Group Collaboration

A paradox lies at the heart of breakthrough research: the tension between individual creativity and group collaboration. Like two sides of a coin, they are inextricably linked and indispensable. Too much emphasis on one may stifle the fruitful potential of the other. Striking the right balance is critical for driving innovative ideas and transforming them into groundbreaking discoveries. This delicate dance is further complicated by the fact that creativity is a rare, intangible quality that transcends mere talent or learned technique. It is the spark that ignites the fire of invention, and like fire, it needs the right conditions to burn brightly and consistently.

Consider the famous rivalry between Thomas Edison and Nikola Tesla - two icons of inventive genius on opposite ends of the creativity spectrum. Edison, the empirical inventor, worked with an army of assistants in his carefully organized workshop, churning out countless inventions through trial and error, guided by the dictum "genius is one percent inspiration, ninety-nine percent perspiration." In contrast, Tesla, the intuitive visionary, designed his inventions in solitude and almost entirely inside his head, relying on remarkable powers of imagination and visualization to solve complex engineering problems without recourse to physical models or prototype testing.

Despite their stark differences in work style, Edison and Tesla shared a few key traits. Both were passionately curious, independent thinkers who cultivated wide-ranging knowledge beyond their primary domain of electrical engineering. In their quest for innovation, they displayed incredible persistence, doggedly pursuing solutions to problems that others had deemed intractable or insoluble. Above all, their legendary accomplishments relied on a delicate balance between individual creativity and orchestration, marshaling the human and material resources needed to bring their ideas to fruition.

One way to maintain this balance is by creating an environment that encourages a free flow of ideas and allows people to take risks without fear of judgment or reprisal. The concept of psychological safety, first introduced by Harvard professor Amy Edmondson, is one of the key factors when it comes to fostering a culture of innovation within a group or organization. In psychologically safe environments, individuals feel comfortable sharing their thoughts and opinions, which can lead to divergent thinking and the generation of novel ideas. Moreover, healthy debate and constructive criticism ensure that ideas are scrutinized, refined, and improved upon before they reach the implementation stage.

Another aspect to consider in this delicate equilibrium is the relationship between homogeneity and heterogeneity within the group. While cohesion and a shared identity can promote trust and facilitate communication, too much similarity can lead to stagnation and groupthink. This is where cognitive diversity becomes essential. By bringing together individuals with contrasting perspectives, expertise, and styles of thinking, the group can collectively produce more innovative solutions to complex problems, drawing upon a richer pool of ideas and avoiding the pitfalls of insularity.

One particularly striking example of the successful balance between individual creativity and group collaboration can be found in the story of the Manhattan Project. At the height of World War II, brilliant scientists and engineers from diverse backgrounds, cultures, and disciplines were assembled under the pressure-cooker conditions of a race against time to create the first nuclear weapons. The intense intellectual environment of the project, combined with the unprecedented scale of collaboration and knowledge-sharing, ultimately gave birth to the atomic era. Out of this crucible emerged both transformative research and extraordinary tales of personal growth and self-discovery.

Even outside the rarefied world of scientific research, striking the right balance between individual creativity and group collaboration remains a universal challenge. Anyone who has participated in a brainstorming session, a writer's workshop, or coordinated a group project can attest to the potential of combining creative minds for achieving something greater than the sum of its parts. However, an effective framework of balance must also consider the human tendency towards social comparisons, displaying a blend of humility and pride, allowing compliments and criticism to coexist.

In navigating the intricate dance between individual creativity and group collaboration, modern research endeavors can learn invaluable lessons from the pioneers who came before: Edison and Tesla as paragons of inventive genius, the Manhattan Project as a testament to the power of collaboration in a complex, high-stakes environment. By embracing cognitive diversity, cultivating psychological safety, and nurturing the right balance between autonomy and teamwork, we can create powerful synergies that fuel innovation and drive the progress of human knowledge to new, as yet undreamt, frontiers of inquiry.

Strategies for Encouraging Healthy Collaboration Within Research Teams

A key component of encouraging healthy collaboration is creating an environment that fosters intellectual curiosity, risk-taking, and mutual respect.

One way to achieve this is to adopt a questioning and learning culture, where individuals feel comfortable asking for help, challenging assumptions, and proposing alternative solutions to problems. For instance, Charles Darwin often engaged his colleagues in spirited debates on the mechanisms of natural selection, thereby leading to a more robust understanding of evolutionary theory. In a research group culture that values open communication and active listening, individuals learn from their peers and collectively forge a deeper understanding of the problem they are investigating.

Another element of healthy collaboration is the cultivation of trust and psychological safety within the team. This can be achieved through regular check-ins, where team members discuss not only their project's progress but also the process and the interpersonal dynamics in the group. Such open discussions can reveal potential issues related to interpersonal conflict, lack of clarity regarding roles, or differences in working styles. Importantly, these discussions should not be seen as a source of blame or criticism but as an opportunity for constructive feedback and growth. By fostering trust and psychological safety, research groups can create a supportive environment that allows individuals to push through inevitable setbacks and failures, leading to more innovative and resilient solutions.

Diversity of thought and expertise is vital for meaningful collaboration, as it brings forth novel insights and perspectives. Teams must recognize and nurture this diversity to ensure that all members feel valued and heard. For example, the late cognitive psychologist Amos Tversky and economist Daniel Kahneman challenged conventional wisdom in their respective fields by taking an interdisciplinary approach to decision-making, resulting in groundbreaking work on cognitive biases and heuristics. By respecting and embracing differences in background, training, and perspective, teams can develop unique approaches to problem-solving that may not have been possible if everyone thought alike.

Reaping the benefits of collaboration also requires clear processes and shared expectations regarding how work is to be conducted. Defining roles and responsibilities can significantly improve team interactions by minimizing confusion and uncertainty regarding each individual's contribution. Furthermore, articulating and agreeing on shared goals and vision for the project will ensure that everyone is aligned and working towards common objectives. By doing this, research teams avoid stagnation and ensure that all members remain motivated and invested in their collective work.

Finally, it is essential for healthy collaboration to recognize and cele-
brate individual accomplishments and milestones. By acknowledging and
reinforcing individual contributions, teams emphasize the importance of
each member's expertise and input, which in turn promotes engagement and
commitment to the group effort. By celebrating progress, research teams
maintain a sense of momentum, which is critical for sustaining long-term
collaborative efforts.

In conclusion, cultivating healthy collaboration within research teams
requires continuous nurturing and fostering of clear communication, trust,
diversity, shared expectations, and acknowledgment of accomplishments.
By incorporating these elements into their interactions, research teams can
unlock the full power of their collective intelligence and generate groundbreaking innovations that no individual could achieve alone. As Galileo once
stated, "You cannot teach a person something they do not already know,
but in a synergistic collaboration, you spark the magic of understanding and
wisdom, igniting the fires of creativity and innovation." By embracing these
strategies, research teams stand at the threshold of discovering a new world
of possibilities, conquering the yet unchartered territories of knowledge, and
together ushering in a new era of human understanding.

Promoting Intellectual Diversity and Constructive De-
bate

Within the halls of the most innovative and impactful research organizations,
you can't help but notice that their workspaces and teams have a particular
quality to them - a buzzing atmosphere and a sense of irrepressible intellectual vitality. This vigor stems from what is at the heart of every scientific
breakthrough or technological revolution - the idea of intellectual diversity
and constructive debate. To create substantial changes and to foster truly
groundbreaking ideas, researchers and scientists from various backgrounds,
expertise, and perspectives need to collaborate fruitfully, challenge one
another's assumptions, and engage in rigorous debate purposefully.

Let's delve into some of the fascinating examples of intellectual diversity
and constructive debate in action, which have propelled a variety of fields
to greater heights. One notable instance is the groundbreaking work of the

Human Genome Project - a massive international collaboration involving scientists, researchers, and funding from numerous countries and countless expertise domains. This herculean undertaking demanded a synthesizing of knowledge and a constant flow of debate, verbal sparring, and intellectual cross - pollination. Through persistence and intellectual cooperation, the completed project revolutionized our understanding of genetics, allowed for advanced medical diagnostics, and paved the way for personalized medicine.

While the Human Genome Project is a prime example of an international collaboration, intellectual diversity doesn't always require a global scale to spark change. Consider the case of the renowned Bell Labs in the 20th century. Bell Labs, known for its labyrinth - like architecture, was deliberately designed to encourage cooperation among various disciplines, stimulating dialogue and provoking frequent encounters between scientists and engineers who would otherwise work in isolated silos. This orchestrated cross - disciplinary mingling resulted in some of the most influential innovations in the history of communication technology, such as the invention of the transistor and the development of long - distance telephone networks. The thriving intellectual climate of Bell Labs demonstrates that simply changing a physical environment can greatly increase the level of interaction between differently oriented minds, leading to a heightened degree of collaboration and ingenuity.

However, promoting intellectual diversity and constructive debate requires more than just a purposeful mix of talents and expertise. It also depends on creating a culture that encourages vulnerability and risk - taking, allowing creative friction to surface. This can be exemplified by the legendary Xerox PARC research center, which had a profound impact on the digital revolution by nurturing inventions such as the graphical user interface and the Ethernet. What made Xerox PARC so exceptional was its unshakable commitment to fostering a culture of open inquiry, curiosity, and intellectual fearlessness - an environment where researchers could openly challenge one another, explore alternative pathways, and dabble in seemingly heretical thinking without fearing the reprisal of the status quo.

Of course, not all scientific debates are successful, and some can even become contentious and counterproductive, especially when they devolve into personal conflicts. The key to ensuring that debates remain constructive is to maintain a level of intellectual humility, respect, and openness towards alternative viewpoints. Great researchers and scientists can pursue their own agendas with tenacity and vigor, yet still submit to the higher principles of scientific inquiry - always being ready to admit the limitations of their own knowledge and to update their beliefs in the face of new evidence or perspective.

One way to foster these intellectual virtues and encourage constructive debates in research settings is to implement a simple practice known as the "premortem." Before embarking on a high - stakes project or making crucial decisions, the team members spend time collectively imagining why that project might eventually fail, exposing all the potential risks, blind spots, assumptions, and weaknesses lurking beneath the surface. By adopting a spirit of ironic detachment and venturing outside their entrenched convictions, researchers can mitigate confirmation bias and spark meaningful and insightful discussions that change the very trajectory of their work.

As the curtain falls on these historical scenes of intellectual diversity and constructive debate, we must internalize the lessons offered and ask ourselves how to apply this wisdom to our own research endeavors. We should take the time to reflect on our environment, our culture, and our habits - seeking ways to foster vulnerability, collaboration, and curiosity at every opportunity. By embracing these principles, we join an illustrious lineage of pioneers who have dared to challenge one another and ultimately changed the world. And as we forge ahead into previously uncharted territories, we find ourselves further intertwining our intellects with those of our peers in pursuit of the next groundbreaking breakthrough and greater human understanding.

Navigating Conflicting Ideas and Interests to Achieve Collective Goals

Navigating conflicting ideas and interests is an art that must be mastered by researchers, innovators, and visionaries. Picture this: Tesla and Edison in a room, arguing over the merits of alternating versus direct current electricity. Copernicus, clashing with the entrenched belief in the geocentric model of the universe. Francis Crick and James Watson, sparring gracefully in their pursuit of unraveling the mysteries of DNA helix. What separates these historical figures from others not only includes their brilliance and passion but also their capacity to navigate conflicting ideas and interests in pursuit of collective goals.

One of the remarkable examples in recent history highlighting the fine navigation of conflicting ideas to achieve collective goals is the collaboration between competition giants Apple and Microsoft. Although both companies had fundamentally different visions for their products, they recognized a mutual interest in each other's success. In August 1997, Microsoft invested $150 million in Apple to help save the struggling company. In return, Apple agreed to make Microsoft's Internet Explorer its default browser on Mac computers. This alliance not only brought competing product developers together, but it also fostered innovation for both companies.

The success of collaborative research also hinged upon the ability to navigate conflicting ideas and interests. In 2012, Elon Musk's Space X teamed up with NASA to develop reusable spacecraft for space missions. Although both organizations pursued different goals and operated with separate infrastructures, they recognized the mutual advantages that could stem from collaboration. Ultimately, this resulted in SpaceX successfully launching and landing Falcon 9, a game-changing development in spaceflight.

To traverse the critical labyrinth of conflicting ideas and interests successfully, consider applying these principles:

1. Seek Understanding: Strive to not just to hear the opposing viewpoint but genuinely understand its perspectives, motivations, and merits. Engage in active listening and open-minded questioning. By doing so, you create a foundation for mutual respect and dialogue necessary for surfacing innovative solutions.

2. Find Common Ground: Identify the shared values and goals that underpin differing ideas and interests. In the case of Apple and Microsoft, it was the mutual desire for a competitive and flourishing technology market. Establishing common ground will help anchor collaborative discussion and enhance cooperation.

3. Reframe the Debate: When conflicting ideas and interests seem insurmountable, transform the discourse. Explore alternative frames, challenge existing assumptions, and consider new perspectives. Encourage your team to broaden their mental horizons and, in doing so, shatter the barriers that inhibit collaboration.

4. Create a Risk-Tolerant Environment: Embolden your team to share unorthodox ideas and challenge conventional wisdom. In a climate of

psychological safety, risk-taking, and encouragement, collaboration thrives, and creative solutions emerge.

5. Model Respectful Discourse: Lay the groundwork for a culture of respectful debate and disagreement. Recognize that conflict, when handled productively, can be an engine for collective growth.

As the lights slowly dim on the contentious stage of Tesla and Edison, imagine this vision instead: a new generation of innovators, trained in the art of navigating conflicting ideas and interests, coming together to tackle the grand challenges of our time. Their ability to bridge divides, engage in collaborative inquiry, and reap the benefits of constructive conflict not only leads to breakthrough discoveries but also propels humanity forward.

Just as Apple and Microsoft once joined forces, or SpaceX and NASA found collaboration in the pursuit of a groundbreaking enterprise, so too may the next generation of innovators learn to navigate the complex currents of ideas and interests, reaching collective goals that propel humanity ever forward in our inexorable journey of discovery and achievement.

Case Studies of Successful Collaborations and their Impact on Breakthrough Research

One such collaboration that revolutionized the course of history is that of Watson and Crick in their quest to unravel the structure of DNA. In the early 1950s, these two young scientists from different disciplines - one trained in physics, the other in biology - joined forces to tackle an enigma that had baffled researchers for decades. Through their teamwork, complemented by the invaluable contributions of Rosalind Franklin and Maurice Wilkins, they pieced together the puzzle of the DNA double helix, thereby unlocking the genetic blueprint of life itself. This watershed moment heralded the era of modern molecular biology and paved the way for groundbreaking advancements in fields like genetics, bioinformatics, and personalized medicine.

Another example of symbiotic collaboration is the Manhattan Project - the top-secret initiative that brought together some of the finest scientific minds of the 20th century, including Oppenheimer, Fermi, and Szilard. Their mission was to invent the world's first atomic bomb before Nazi Germany could do the same. Despite coming from diverse intellectual backgrounds, these visionaries coalesced into a formidable team that transformed scientific research and demonstrated the power of human innovation - albeit through a devastating instrument of warfare. Remarkably, aside from forging new ground in nuclear physics, the Manhattan Project catalyzed the adoption of interdisciplinary collaboration as a crucial tool for propelling future scientific discoveries.

In more recent times, the Human Genome Project serves as an emblem of synergy and cooperation, uniting researchers from various countries, institutions, and disciplines in the grand mission to decode the human genetic code. In tackling this colossal undertaking, scientists worked collaboratively to develop new techniques and technologies, share data and resources, and overcome countless obstacles - all in the pursuit of a single, unified objective. The fruits of their labor now serve as the cornerstone for countless biomedical research efforts that seek to improve human health and reduce the burden of disease on a global scale.

Similarly, the Large Hadron Collider (LHC) exemplifies the power of international collaboration, bringing together thousands of top-tier physicists and engineers from over 100 countries with the aim of probing the fundamental constituents of our universe. Through their combined expertise and effort, these researchers uncovered the elusive Higgs boson in 2012 - a subatomic particle whose existence confirms the theoretical framework that underpins our understanding of how particles acquire mass. This monumental achievement underscores the importance of uniting minds across geographic, cultural, and disciplinary divides to advance the frontiers of knowledge for the betterment of humankind.

These case studies illuminate a vital message: When we collaborate, we are able to transcend the limitations of individual expertise and see the potential interconnections that form the fabric of the cosmos. Such partnerships create a fertile ground for cross-pollination of ideas, from which the seeds of inspiration and creativity may sprout.

It is through these and countless other examples of successful collaborations that we glean the importance of fostering an environment that encourages intellectual diversity, open communication, and mutual respect. In doing so, we cultivate the conditions necessary for a more profound understanding of our world and ourselves. As we proceed through the uncharted realms of scientific discovery, let us imbue our research endeavors with the spirit of such visionary collaborations, knowing that it is not only our knowledge and skills but also our unswerving commitment to working in harmony that will propel us into a future of untold possibilities and unimaginable breakthroughs.

Chapter 8

Perseverance and Inspiration: Learning Resilience and Rituals from Franklin, Curie, and Darwin

The flame of perseverance burnt strong in the heart of Benjamin Franklin - polymath, founding father, and a seminal member of the scientific revolution. His contributions to the field of electricity were groundbreaking and largely driven by his relentless pursuit of knowledge. Franklin's grit and determination can be best understood by his now-famous kite experiment, conducted during a thunderstorm with an iron key tied to the bottom of a wet kite string. Foolhardy as it may have been, the success of the experiment lay in Franklin's resolute desire to seek answers, even in the face of mortal danger. This unyielding spirit characterized his approach to scientific inquiry and has served as fodder for generations of researchers to come.

Marie Curie stands as another paragon of indefatigable perseverance. Born in Poland, she overcame financial and cultural constraints to become the first woman to win a Nobel Prize, and the only person to have won in two distinct scientific fields - Physics and Chemistry. Curie embodied a pioneering spirit, venturing boldly into uncharted territories of scientific exploration. Of particular note are her painstaking efforts in isolating radioactive isotopes; such was her drive for discovery that when faced with the lack of a suitable workspace, she repurposed an abandoned shed into a makeshift laboratory and soldiered on with her experiments. This unwavering resilience in the pursuit of knowledge continues to be an inspiration in the annals of scientific advancement.

Similarly, Charles Darwin's voyage aboard the HMS Beagle brought about a monumental shift in our understanding of the natural world. His theory of natural selection, encapsulated in the groundbreaking work "On The Origin of Species," laid the foundation for the modern field of evolutionary biology. Darwin's five - year - long circumnavigation of the globe was rife with innumerable challenges; from seasickness and precarious living conditions to the mental turmoil of grappling with the implications of his nascent theory. Nonetheless, his insatiable curiosity drove him to persist in his quest for knowledge and allowed him to overcome adversity with fortitude, ultimately yielding transformative discoveries that continue to shape the trajectory of scientific knowledge.

These three eminent figures possessed unique qualities that encouraged resilience and served as wellsprings of inspiration. Franklin was deeply committed to disciplined self - improvement, instituting a rigorous system of goal - setting and habit - building to maintain a clear, focused perspective on his work. Curie's relentless pursuit of scientific knowledge demanded that she embrace a daily routine marked by unwavering dedication and single - minded focus, allowing her to achieve groundbreaking insights. Meanwhile, Darwin's long and arduous journey on the Beagle fostered a deep reservoir of patience and persistence in his quest to unlock the secrets of the natural world.

For the aspiring researcher, there exists a wealth of wisdom to be gleaned from the lives of these three exemplary individuals. The core message at the heart of their legacies is clear: an undaunted spirit of perseverance and commitment to a disciplined routine can yield wonders in the search for scientific understanding. By embracing this spirit and incorporating elements of Franklin, Curie, and Darwin's methodologies into one's personal research toolkit, the modern - day investigator can traverse the often winding paths of discovery with newfound resilience and a greater aptitude for uncovering the hidden truths that lie in wait.

Building Resilience: Embracing Adversity as an Opportunity for Growth

Consider Sir Isaac Newton, who, during a period of social isolation due to the Great Plague in 1665, made some of his most profound discoveries in mathematics and physics while confined to his childhood home. Removed from the social distractions of academic life, he honed his focus and produced transformative works such as the development of calculus, the formulation of the laws of motion, and the understanding of universal gravitation. It was in this crucible of adversity that Newton's true genius began to emerge. Recognizing the potential positive aspects of such hardship can be a valuable cognitive reframing technique to foster resilience in research.

Similarly, the life and work of Albert Einstein offer ample evidence of the importance of resilience in overcoming obstacles to scientific progress. Despite encountering numerous setbacks, including failing his initial entrance exam to the Swiss Federal Institute of Technology and being rejected for multiple teaching positions, Einstein persevered. He found work as a patent clerk and continued refining his theoretical ideas in his spare time. Ultimately, his determination culminated in the publication of four groundbreaking papers in 1905, forever changing our understanding of the universe through his theories of special relativity, the photoelectric effect, and the formulation of the famous equation, E=mc. Einstein's tale serves as a potent reminder of the transformative power of relentless commitment to one's passion in the face of adversity.

When embarking on the path of scientific inquiry and research, we should also glean inspiration from Thomas Edison, whose resilience in the face of failure is the stuff of legend. His seemingly endless series of experiments with the electric light bulb and other inventions tested numerous potential avenues, often with no initial success. Edison's famous quote - "I have not failed. I've just found 10,000 ways that won't work" - encapsulates the mindset of framing adversity as an opportunity for growth. By embracing the notion that each failed experiment brought him one step closer to a solution, he built the resilience to press forward and ultimately revolutionized the world with his inventions.

How can researchers cultivate this invaluable quality of resilience? One crucial aspect is to adopt a growth mindset, as described by psychologist

Carol Dweck. Instead of viewing intelligence and talent as fixed traits, a growth mindset emphasizes the development of abilities through dedication and hard work. By understanding that failing or encountering roadblocks is a natural part of the learning process, researchers can better adapt to challenges and persist in seeking novel solutions to complex problems.

Moreover, embracing the serendipitous nature of discovery can bolster resilience. Unanticipated obstacles often reveal unexpected insights and potential paths for further exploration. Viewing setbacks as opportunities for serendipity can provide motivation to continue probing the depths of scientific knowledge. The discovery of penicillin by Alexander Fleming offers an instructive example of this concept in action; a seemingly mundane bacterial contamination of his petri dishes led to the accidental discovery of the groundbreaking antibiotic, changing the course of medicine for generations to come.

In the pursuit of scientific breakthroughs, adversity becomes an inseparable companion. Entwined with curiosity and persistence, it forms the threads that weave the tapestry of knowledge through which humanity progresses. As each visionary researcher and scientist embraces adversity and builds resilience, they strengthen not only their resolve and determination but also the very fabric of the scientific endeavor, allowing the legacy of innovation to continue.

As we delve further into the inner workings of visionary minds and how their ideas have shaped our understanding of the world, we must appreciate the power of resilience in overcoming cognitive biases, fortifying objectivity, and embracing the beginner's mind - aspects that refine and enrich the process of scientific inquiry and discovery. Failure, far from being an end point, is but a stepping stone on the path to transcendent achievement, inviting those brave enough to embrace adversity as an opportunity for growth.

Benjamin Franklin's Model of Perseverance: Lessons from a Life of Grit and Discipline

Benjamin Franklin, polymath and one of America's founding fathers, is known for his inexhaustible energy and tireless pursuit of knowledge. Throughout his life, Franklin displayed an unwavering commitment to self-improvement, using grit and discipline as driving forces behind his numerous achievements in science, writing, and diplomacy. Studying his life offers countless lessons in perseverance, which can be applied to strengthen one's approach to scientific research today. By examining the four key components of Franklin's model of perseverance - goal - setting, planning, the art of self - reflection, and adaptive learning - we can learn how to cultivate greater resilience and determination in our pursuit of truth and discovery.

First and foremost, Franklin was a master of goal - setting. He strategically identified objectives that he deemed essential for self-improvement and used these goals to guide his actions. For instance, he devised his famous "13 Virtues" plan - a meticulous list of moral qualities that he believed would lead to personal success. Each week, Franklin would dedicate himself to practicing one of these virtues, recording his daily progress in a notebook. By setting clear and attainable goals for himself, Franklin created a strong foundation for personal growth, which in turn fueled his perseverance and motivated him to tackle increasingly ambitious challenges.

In the realm of scientific research, setting clear and attainable goals is an invaluable skill for fostering perseverance. By breaking down complex problems into smaller, manageable objectives, researchers can create a clear roadmap to success and maintain a sense of progress along the way. Moreover, goal - setting can help refine a researcher's focus, ensuring that they invest time and energy into the most critical aspects of their work and avoiding the tendency to overextend themselves.

Franklin's second pillar of perseverance was an unwavering dedication to planning. In his Autobiography, he famously laid out a detailed account of his daily schedule. This structured approach to time management enabled him to allocate specific hours for work, study, reflection, and relaxation, bringing balance and order to his life. By adhering to a well - honed routine, Franklin was able to maintain peak productivity and prevent burnout - a vital trait for any individual engaged in the pursuit of knowledge.

For modern researchers, adopting a structured and disciplined approach to time management can be transformative. By carving out designated time for deep, focused work, and ensuring regular breaks for physical and mental rejuvenation, scientists and academics can maintain peak performance while minimizing the risk of burnout or mental fatigue.

Thirdly, Franklin was a staunch advocate of self - reflection. He regularly

assessed his performance against his goals and sought to understand how
he could further improve. This process of introspection helped him to
identify weaknesses and the areas of his life that required greater discipline or
determination. By fostering a continuous cycle of reflection and improvement,
Franklin stayed motivated, despite setbacks or failures.

In the context of scientific research, cultivating the habit of self-reflection
can lead to profound insights and improvements in one's work. By regularly
evaluating the progress of a research project, asking critical questions,
and assessing the effectiveness of one's methodology, researchers can make
informed adjustments as necessary, paving the way for more robust and
meaningful findings.

Franklin's final component of perseverance centered on adaptive learning.
He embraced failure as an opportunity for growth and viewed setbacks as
a natural part of the learning process. This resilient mindset enabled him
to continually innovate and push the boundaries of his knowledge. For
example, when Franklin encountered skepticism and doubt in his pursuit
of the practical applications of electricity, he turned to experimentation,
confidently and systematically debunking the misconceptions of his time.

Franklin's adaptive learning approach is indispensable for modern researchers, who must often navigate difficult questions and uncertainty in
their work. By embracing failure as a learning experience rather than a
mark against their abilities, scientists can foster the resilience necessary to
persist in the face of adversity.

In examining the life of Benjamin Franklin, we uncover a powerful model
of perseverance, characterized by goal-setting, planning, self-reflection,
and adaptive learning. His disciplined approach to self-improvement and
determination in the face of adversity provide valuable lessons for researchers
embarking on their journey for scientific truth. By applying these principles
to their own work, scientists and academics can embark on a journey not
unlike Franklin's own: a life defined by grit, discipline, and a relentless
pursuit of knowledge. Indeed, the spirit of Franklin lives on in those who
carry the torch of discovery forward, illuminating the path to truth one step
at a time.

Marie Curie's Rituals for Daily Inspiration: Pioneering Creativity through Routine and Hard Work

Marie Curie, a name etched in the annals of history and synonymous with the radiant glow of her groundbreaking discoveries, was not merely a brilliant scientist but a true pioneer in using daily rituals to fuel her unwavering determination, creativity, and hard work. From the depths of her somber childhood in partitioned Poland to her noble ascent into the pantheon of scientific legends, Marie Curie's inspirational life story is permeated with instances of her incredible work ethic and the hallowed routines she meticulously adhered to, which have served as invaluable lessons for the generations that have followed in her footsteps.

The days of Marie Curie's youth were far from idyllic, with the loss of her mother and sister plaguing the young Marie with a sorrowful burden. Despite her family's meager financial means, Marie and her siblings were driven by a hunger for knowledge, largely due to their intensely intellectual father who instilled in them the value of education. Fueled by the immense challenges she faced right from the outset, Marie's commitment to hard work and a diligent regimen was forged, forming the bedrock upon which she would construct her legacy.

Studying in secret rendezvous known as the "Floating University" in a time when the Russian authorities were stifling Polish academics, Marie Curie's thirst for knowledge could not be quenched by the oppressive regime, and she used these trying times to create her own system of learning and her unique approach to discovery. After a period of working to support her sister's medical studies, Marie left Poland and pursued higher education in Paris at the prestigious Sorbonne University, where her unwavering focus and innate curiosity became the linchpin of her future success.

With a strong foundation in research and an appetite for knowledge, it was not long before Marie rose to prominence and, with her husband Pierre Curie, made significant inroads into the mysteries of radioactivity. During their days of relentless experimentation, the Curies and their insatiable love for investigation remained steady and true, with Marie Curie's tireless daily rituals acting as the constant spark that ignited their shared passion.

Marie Curie's daily rituals consisted of arduous hours in the laboratory, replete with physical labor and the perils of handling hazardous materials.

Often clad in dusty, old clothing- to the point where fellow scientists would refer to her as a "beggar"- Curie spent countless hours in solitary toil, demonstrating a fierce determination to push the boundaries of scientific discovery. Through these hours of grueling work, Marie Curie was able not only to isolate the new elements of polonium and radium but to dedicate herself to the application of these elements in the betterment of humanity.

One particular example that stood out in the annals of Curie's remarkable life was her utilization of her own daily habits to engage in an act of pure selflessness. During World War I, Curie repurposed her knowledge of radium and her skills in mobile X-ray technology to create ambulances with built-in X-ray equipment, thus earning her the nickname of "Little Curie." Striving to alleviate the suffering of war-wounded soldiers, Marie Curie put her own health at risk, bearing the brunt of her unbounded commitment to humanity.

The tale of Marie Curie's life, scattered amongst the pages of scientific history, is a testament to the power of unwavering commitment, perseverance, and the potential for greatness that lies within each person who undertakes the laborious journey of research. But most importantly, Marie Curie's story serves as an illumination of the profound impact that daily rituals- the sacred, the mundane, and everything in between- can have upon the ways in which creativity and a pioneering spirit can be harnessed and wielded in the noble pursuit of knowledge.

While Marie Curie's exceptional intellect underpins her legendary status, it is the daily habits she upheld- her discipline, her dedication to learning and intervening in a world beset by suffering, her willingness to weather adversity - that make her an indelible beacon of inspiration to researchers across all disciplines.

As we strive to emulate the spirit of Marie Curie's tenacious perseverance and creativity, let us bear in mind that pursuing a bold new world of discovery begins by fostering the discipline to approach our research with focused, unwavering resolve. Let it be our guiding light as we seek to understand the complex connections between myriad disciplines, from biology to art and far beyond. And let it lead us onward in probing the depths of our own curiosity, forging new paths, and making history of our own.

Charles Darwin's Journey of Discovery: Cultivating Persistence and Patience in Research

Delving into the extraordinary voyage of Charles Darwin on the HMS Beagle, we explore how his exceptional persistence and patience laid the foundation for one of the most groundbreaking theories in modern science: the theory of evolution through natural selection. As we embark upon this tale, we discover how Darwin's qualities manifested in his journeys, observations, and the development of his revolutionary ideas.

In 1831, a young Charles Darwin stepped aboard the HMS Beagle as the ship's naturalist. Over the next five years, the Beagle's journey took him across the world, from the coasts of South America to remote islands in the Pacific. This incredible journey provided Darwin with the opportunity to meticulously observe and document various species of flora and fauna, ultimately culminating in the formulation of his groundbreaking theory. With no modern - day conveniences or guarantees of success, Darwin's determination and patience were put to the ultimate test.

As Darwin set sail on this voyage of discovery, he faced numerous adversities. The harsh weather and unforgiving seas pushed the young naturalist to the brink of physical exhaustion, while his seasickness and homesickness took an emotional toll on him. However, the ultimate measure of Darwin's impressive perseverance was his unwavering commitment to his research even in the face of these adversities.

During his travels, Darwin maintained a rigorous schedule of specimen collection, detailed documentation, and analysis. With determination and focus, he recorded precise notes on each specimen encountered, from the native plants and animals to fossils and geological formations. It was his insatiable curiosity and a deep understanding of the importance of systematic data collection that allowed him to endure the challenging conditions of his journey.

Darwin's patience was not only evident in his painstaking work during the voyage but also during the years of research that followed his return to England. He spent over two decades rigorously analyzing and building on the observations and data he had accrued during his voyage on the Beagle. As he painstakingly dissected, categorized, and compared the various species he had documented, Darwin began to formulate his momentous ideas that would later be compiled in "On the Origin of Species."

Even as the evidence in favor of his proposed natural selection theory accumulated, Darwin exhibited an unwavering commitment to scientific integrity and intellectual honesty. Instead of hastily announcing his ideas, he meticulously refined his theory, ensuring that it was built upon a solid foundation of empirical evidence.

In a remarkable display of patience and diligence, Darwin withstood challenges and criticisms while gradually and carefully developing his theory, without succumbing to the temptation of premature disclosure. As he corresponded with fellow scientists and navigated his way through the treacherous landscape of 19th-century scientific debate, his determination and measured approach proved instrumental in establishing the credibility of his revolutionary ideas.

In the wake of Darwin's groundbreaking achievements, we can learn valuable lessons about the impact of persistence and patience in the world of scientific research. The tireless efforts and unwavering commitment to gather and analyze information are qualities that all researchers must strive to embody in order to make similarly groundbreaking discoveries.

As we progress through our own scientific journeys, we can draw inspiration from Darwin's voyage aboard the Beagle and his enduring legacy. His exemplary dedication to rigorous examination and bearing the tests of time in an ever-consuming quest for truth demonstrates the importance of resilience in the face of adversity, obstacles, and uncertainties that any researcher may encounter.

When we allow ourselves to embrace the lessons from Darwin's exceptional journey, we are better equipped to undertake our own ventures into the vast, uncharted territories of scientific discovery. With each challenge faced, we can remember that, much like Darwin, our own patience and persistence could one day unveil truths that shape the future of humanity. And as we, too, look upon the natural world with a sense of wonder and meticulously record the intricacies of our observations, we likewise become agents of awe-inspiring progress and collective enlightenment.

Fostering a Mindset of Perseverance and Inspiration: Strategies to Overcome Obstacles and Stay Motivated

Innovation and groundbreaking discoveries are propelled by individuals who maintain unwavering concentration and motivation amid uncertainty and adversity. One can glean valuable insight on nurturing a mindset of perseverance and inspiration through examining the lives of great thinkers and inventors. By understanding the obstacles they faced, and the strategies they used to overcome them, we can decode the secret formula to enduring motivation and persistence in research.

Perseverance and inspiration come from within, and the strength of the human spirit is visible in countless examples throughout history. Charles Darwin, the celebrated English naturalist and geologist, was often plagued by doubts and faced considerable backlash from his contemporaries. His journey to success began as a young, inexperienced naturalist aboard the HMS Beagle. Darwin spent years meticulously accumulating information on various plant and animal species, which eventually formed the backbone of his groundbreaking theory on evolution. The secret to his motivation was a deep-rooted passion for knowledge and a profound sense of curiosity that powered him through moments of doubt and criticism. Opening oneself up to renewed curiosity and an insatiable appetite to learn can fuel unprecedented perseverance and inspiration.

Another example of resilience in the face of adversity is the story of physicist and chemist, Marie Curie. Her work was conducted during a time when women faced extreme discrimination in the scientific community. Undeterred, she became the first woman to win a Nobel Prize and remains the only person to have claimed the award in two separate scientific disciplines: physics and chemistry. Despite facing numerous adversities - losing her husband and research partner, facing criticism from male colleagues, and confronting health issues due to prolonged exposure to radiation - Curie's devotion to her work never wavered. One of the reasons for her unwavering dedication was her meticulous approach to documentation and organization of her research, which helped her stay focused and motivated despite the setbacks. The power of routine and organizing one's work to foster motivation and clarity cannot be underestimated.

Nikola Tesla, the inventive engineer and physicist, is another example of unyielding motivation amidst numerous obstacles. He made a tremendous impact on the world of electricity and holds more than 300 patents to his name. Tesla's determination to bring his ideas to fruition, even in the face of financial hardships, enabled him to have a lasting impression on society. One key aspect of Tesla's mindset was his ability to visualize his ideas, which allowed him to gain a clearer understanding of his objectives and maintain a steadfast focus on his goals. Visualization techniques can be potent tools in cultivating motivation when faced with temporary setbacks.

To foster a mindset of perseverance and inspiration, one must develop coping strategies tailored to individual strengths and weaknesses. Inspired by the lives of famous inventors and pioneers, several approaches can bolster motivation and resilience, such as cultivating curiosity, fine-tuning organizational skills, and practicing visualization. Indulging in passions and taking calculated risks can also lead to more profound dedication and willpower even when faced with challenges.

Ultimately, perseverance and inspiration are the driving forces behind the relentless pursuit of knowledge, overcoming the barriers that hinder progress in research, and challenging conventional wisdom. As trailblazing polymath Leonardo da Vinci once said, "Obstacles cannot crush me. Every obstacle yields to stern resolve." Adopting this resolve, embracing curiosity, refining organizational skills, and harnessing the power of visualization can set researchers on a path to achieving success beyond imagination. Bolstered by this unshakable foundation, one can break the barriers of limitations and embrace the infinite potential of the human mind, just as Tesla, Darwin, and Curie did before them.

Integrating Resilience and Rituals into Your Research Toolkit: Practical Applications for Breakthrough Success

The keys to achieving breakthrough successes in research often lie in the balance between resilience and well-developed rituals that researchers integrate into their lives and work. To better understand this, we will look at how we can integrate resilience and rituals into our research toolkit, with practical examples and strategies to ensure remarkable outcomes.

Resilience is the ability to withstand adversity, bounce back, and transform challenges into opportunities for growth. In the context of a researcher, resilience is essential to navigate the ever-evolving landscape of discovery and setbacks. As Thomas Edison famously said, "I have not failed. I've just found 10,000 ways that won't work." By embracing this mindset, researchers can transform failures into valuable lessons regarding which strategies to discard and which to adopt.

Practical applications to building resilience in research include setting realistic expectations and being prepared for setbacks. Creating an environment where failure is not feared, but rather seen as a stepping stone to a new solution, can foster adaptation and courage. Moreover, developing strong coping skills by seeking mentorship, learning from colleagues, or attending workshops can bolster resilience.

On the other hand, rituals in research can take the form of routines, habits, and mental exercises that serve to refine one's intellect and stimulate curiosity. Daily rituals can create an anchoring effect, providing stability and comfort in the face of uncertainties and setbacks that often permeate the world of research. Some rituals might include regular breaks for reflection or mindfulness practices, designating specific times of a day for exploration or absorbing information, committing to a weekly review of goals, and engaging in activities that replenish mental and emotional resources.

Here are some examples and practical applications to integrate resilience and rituals into your research toolkit for breakthrough success:

1. Adapt a growth mindset: Recognize that the path to discovery is nonlinear and often riddled with obstacles. A growth mindset encourages embracing challenges and seeking continuous improvement. Learn from the setbacks by analyzing what went wrong and what could be done differently next time. Reflect on processes and outcomes regularly to refine approaches and update beliefs.

2. Establish a failure analysis method: When facing a setback or failure, turn it into a learning opportunity by conducting a root cause analysis and documenting the findings. Use this knowledge to guide future research decisions and to develop contingency plans for potential issues that may arise.

3. Develop a daily ritual for idea generation: Set aside time each day for brainstorming and innovation. Treat this time as sacred and commit to it unwaveringly. Additionally, consider engaging in practices that foster free-flowing thought, such as meditation or journaling, to enable structured and

disciplined creativity.

4. Strive for balance: Integrate work - life balance rituals by setting boundaries between professional and personal life. Regularly engage in activities that bring joy and offer a break from the demands of research, such as spending time with loved ones or engaging in hobbies. This balance can offer fresh perspectives and renewed energy when tackling complex research challenges.

5. Cultivate a support network: Develop relationships with mentors, colleagues, and others in the scientific community. Leverage these connections to seek guidance, feedback, and encouragement for ongoing research endeavors. Attend conferences and workshops to establish new connections and learn from the experiences of other researchers.

In observing the lives of luminaries such as Benjamin Franklin and Marie Curie, we can appreciate the profound impact that rituals and resilience have had on their extraordinary careers. A thoughtful cultivation of these elements can lead us not only towards groundbreaking research but also towards a richer and more fulfilled life.

As we journey further into the vast frontier of discovery, the integration of resilience and rituals will become our guiding compass, steering us through the storms of setbacks, providing calm in the face of uncertainty and illuminating the path to breakthrough successes. The researcher who finds the balance between these two seemingly disparate forces will unlock the potential to transform not only their work but also the world around us, impacting the very fabric of human knowledge for generations to come.

Chapter 9

Grace Hopper, Katherine Johnson, and the Principles of Leadership: Guiding Teams to the Frontier of Innovations

Grace Hopper and Katherine Johnson's achievements have stood the test of time and are counted amongst history's most influential contributors to the fields of computer science and applied mathematics. Hopper and Johnson not only pioneered breakthroughs that redefined their respective disciplines but also displayed key leadership principles that have inspired countless individuals and guided teams in their pursuit of innovation. By examining the lives and careers of these extraordinary women, we can distill their leadership principles and apply these lessons to our own creative pursuits.

One of the most enduring examples of their leadership was their ability to navigate uncertainty with remarkable vision, adaptability, and inclusivity. In the early stages of computer science, Hopper envisaged a world in which programming languages would allow humans to communicate more efficiently with machines. This vision led her to develop the first-ever compiler and ultimately lay the foundation for COBOL, a programming language that revolutionized data processing. Hopper embraced the unknown and iterated on her ideas repeatedly, demonstrating the adaptability and determination necessary to accomplish groundbreaking feats.

Similarly, Katherine Johnson exhibited great vision during her time at NASA. Johnson's commitment to accuracy and precision in her calculations propelled her into becoming the first woman to be officially recognized in an author report at NASA. Johnson's work on calculations for orbital mechanics was instrumental in John Glenn's successful orbit around Earth and the Apollo 11 mission to the Moon. Indeed, these accomplishments were facilitated by Johnson's ability to embrace uncertainty and overcome the boundaries imposed by the societal norms of her time.

Both Hopper and Johnson fostered an environment that encouraged collaboration and inclusivity, another key aspect of their leadership. They recognized that fostering a diverse team of talented individuals with different backgrounds and perspectives would be integral to the progress of their respective fields. In fact, they did not shy away from putting themselves forward and making space for other women and members of marginalized groups, understanding the importance of empowering others in bringing about positive change.

For instance, Grace Hopper would make a point of attending Computer Science conferences and actively engage with young professionals, motivating them to remain curious, ask questions, and take risks. She showed genuine interest in their work, providing invaluable guidance that helped shape countless careers. Katherine Johnson, through her efforts to ensure that her team members received appropriate credit for their work, sent a powerful message about the importance of acknowledging and valuing diverse contributions. Together, these actions sent ripple effects throughout their organizations, cultivating a culture of curiosity and openness to new ideas.

In their pursuit of excellence, Hopper and Johnson exhibited an unwavering commitment to developing and nurturing talent. They understood that the key to innovation lies not merely in the discovery of new knowledge but also in fostering the potential of those around them. Through their mentorship, they instilled a lifelong passion for learning and a desire to overcome limitations in the hearts and minds of their protégés.

Grace Hopper created a number of rituals that fostered learning within her teams. She would place a clock on her office wall that ran counter-clockwise as a reminder that 'it is never too late to think differently.' This metaphorical gesture challenged her team to view problems from various angles and to foster innovative ideas. Katherine Johnson, on the other hand, diligently took on the role of mentor to her fellow female mathematicians, sharing essential techniques and strategies that equipped them to excel in their careers.

It is only by examining the individual elements that defined the leadership styles of Grace Hopper and Katherine Johnson that we can begin to understand the true impact they had on their fields and the importance of their approach to leadership in fostering innovation. By embodying these principles - nurturing talent, embracing diversity, and fostering adaptability - we can ensure that we, too, contribute to a legacy that stands the test of time.

As we move forward in our creative journey, let us recognize that these pioneering women continue to be lighthouses, lighting the way to the future for those who dare to embark on their own voyage of discovery. Through their leadership, we are reminded that the true potential for innovation resides within ourselves, just waiting to be unlocked. And with it, the key to redefining the boundaries of human achievement and propelling us ever closer to the frontiers of innovation.

Introduction: The Leadership Legacy of Grace Hopper and Katherine Johnson

Grace Hopper and Katherine Johnson are two formidable figures whose leadership legacies have left an indelible impact on the world of scientific research and innovation. From Hopper's pioneering contributions to computer science to Johnson's groundbreaking work in mathematics and space flight, these trailblazing women demonstrated not only exceptional technical proficiency but also an inspiring ability to lead, mentor, and foster groundbreaking innovations in their respective fields.

At a time when the tech industry was in its nascent stages, Grace Hopper played a pivotal role in birthing modern programming languages. A visionary leader who defied convention, Hopper's strategic thinking and unwavering perseverance led her to develop the first compiler, a tool instrumental in translating human - readable programming languages into machine - executable code - a feat previously thought impossible. As one of the leading minds behind COBOL, one of the earliest high-level programming languages, she played a crucial role in making computer programming more accessible and practical for a wide range of applications.

Hopper's background in mathematics and logic, supplemented by her years of experience as a naval officer, equipped her with a unique perspective that enabled her to unravel complex problems and devise innovative solutions. She championed collaboration and believed that an inclusive, diverse team was essential to uncovering new ideas and forging groundbreaking discoveries. This philosophy was evident in her famous quote, "The most important thing I've accomplished, other than building the compiler, is training young people." By nurturing and mentoring the next generation of computer scientists, Hopper effectively shaped the future of the rapidly growing tech industry.

In a parallel universe of scientific inquiry, Katherine Johnson broke down barriers as an African American woman working at NASA during the height of the Civil Rights Movement in America. Employed as a "human computer," she demonstrated remarkable accuracy and analytical prowess in calculating the trajectories for multiple space missions, including the historic Apollo 11 mission that resulted in the first human moon landing.

Johnson's resolve and unwavering commitment to excellence were evident throughout her illustrious career. She insisted on double-checking her work, scrutinizing the outputs of electronic computers for errors, and rigorously verifying flight calculations. Recognizing the significance of her work, John Glenn, the first American astronaut to orbit the Earth, famously demanded, "Get the girl to check the numbers."

In the face of adversities and challenges that confronted her as a woman and person of color, Johnson remained steadfastly focused on her work, earning the respect and admiration of her peers. She embodied the essence of quiet leadership and unrelenting perseverance in the pursuit of knowledge. Johnson's influence reverberated beyond numbers and equations - she inspired generations of women and minorities to pursue careers in science, technology, engineering, and mathematics, effectively broadening the possibilities for innovation and discovery.

As pioneers in their respective fields, Grace Hopper and Katherine Johnson left indelible marks on the landscape of scientific discovery. They were unyielding in their pursuit of knowledge, undeterred by challenges,

and unwavering in their belief that diverse perspectives and collaboration
were central to unleashing breakthrough ideas. The leadership legacies of
these formidable women invite us to emulate their unquenchable curiosity,
fearlessness in the face of adversity, and commitment to nurturing talent
across generations.

Grace Hopper: A Pioneer in Computer Science and the Birth of Programming Languages

Grace Hopper is a name that conjures images of a dedicated, tenacious, and
visionary computer scientist. Fondly known as "Amazing Grace," her contributions to the world of computing have led to remarkable breakthroughs
that still impact the industry today. But the essence of Hopper's legacy
goes beyond the groundbreaking programming languages she created; it also
lies in her unshakeable determination, intellectual curiosity, and fearless
leadership.

Born in 1906, Hopper was a restless and curious child who harbored
a boundless passion for learning. Her curiosity persisted into adulthood,
leading her to become one of the first women to earn a Ph.D. in mathematics
from Yale University. Torn between joining the mathematics department at
Vassar College and enlisting in the U.S. Navy, she eventually settled on the
latter, setting the stage for her extraordinary career as a computer scientist
and naval officer.

In the early 1950s, Hopper and her team were tasked with developing
a compiler for the first stored program computer, the UNIVAC (Universal
Automatic Computer). Back then, computers could only understand numeric
codes in the binary form, which made programming a laborious and error-
prone endeavor. Driven by her desire to democratize computer programming,
Hopper envisioned a future where anyone, regardless of their mathematical
prowess, could write code in plain language. This vision would mark the
genesis of new, revolutionary programming languages.

United by their shared vision, Hopper's team turned their dream into
reality by creating the first - ever compiler, known as the A - 0. This
groundbreaking invention transformed mathematical symbols into machine
code, freeing programmers from the shackles of binary sequences. Moreover,
the compiler marked the beginning of a new era - the birth of high - level

programming languages.

Building on the A - 0, Hopper spearheaded the development of the revolutionary programming language called FLOW - MATIC. Designed to facilitate efficient information processing for businesses, FLOW - MATIC provided an array of benefits. For the first time in history, programmers could write code in a language that closely resembled ordinary English, opening the doors of the computer programming world to elites and commoners alike.

Given the burgeoning demand for business - oriented software, Hopper's timely invention of FLOW - MATIC had a transformative impact on the computing industry. In fact, FLOW - MATIC's legacy extended beyond its immediate commercial success; it provided the foundation for future computer programming languages, most notably COBOL.

COBOL (Common Business - Oriented Language) is widely regarded as Hopper's crowning achievement. Developed with a consortium of computer manufacturers and leading government agencies, COBOL combined the best features of Hopper's FLOW - MATIC with those of other successful programming languages. COBOL's syntax, characteristically clear and precise, facilitated a seamless translation from human - readable instructions to machine code. Furthermore, COBOL's portability allowed programmers to write a single program that could run on multiple machines, dramatically increasing the efficiency and cost - effectiveness of technological investments.

Grace Hopper's pioneering work in computer programming languages highlights her capability to envision a future beyond the limitations of her present reality. Determined to make programming more accessible, Hopper's unrelenting pursuit of innovation has left an indelible mark on the field of computer science.

Akin to the master painters endowing the world with their masterpieces, Hopper gifted us a myriad of programming languages. Yet, what sets her apart from her peers is her means of inspiration: she painted with algorithms, laid her strokes within the digital realm, and ultimately produced a language fit for all who chose to wield it.

Hopper's legacy can still be felt today, with rippling effects through not only computer programming languages but also the computational world as a whole. Her conviction, determination, and relentless pursuit of innovation continue to empower a new generation of visionaries as they disrupt, create, and ultimately, forge new paths into uncharted realms of discovery.

Katherine Johnson: Breaking Barriers in Mathematics and Space Flight

Katherine Johnson's story surfaces from the depths of history as a bright beacon of hope for generations to come. As a masterful mathematician and physicist, she shattered social and racial barriers, while her relentless determination catalyzed numerous space flight missions and rewrote the course of history. Unlocking Johnson's story paints a vivid canvas, depicting a masterful problem solver, and illuminating the numerous feats she accomplished during her remarkable lifetime.

Born in 1918 in West Virginia, Katherine Johnson showcased a prodigious gift for mathematics from an early age. She was just 18 years old when she graduated summa cum laude from West Virginia State with degrees in mathematics and French. Despite her exceptional talent, her trajectory from there was fraught with hurdles. Segregation and discrimination veiled the landscape, and as an African-American woman, the prospect of a career in research seemed bleak.

Notwithstanding the myriad of restrictions that society imposed upon her, Johnson persisted, and her persistence paid dividends. In 1953, she was hired as a mathematician by the National Advisory Committee for Aeronautics (NACA), the precursor to NASA. Johnson joined an all-black computing group led by Dorothy Vaughan, where she initially held a lowly position as a human computer. However, it wasn't long before her unique aptitude was recognized. Johnson was quickly reassigned to a larger team, which included engineers researching supersonic flight and space exploration.

Johnson played a pivotal behind-the-scenes role in calculating the flight trajectory of Alan Shepard's inaugural manned space flight in 1961. To put this into context, consider that in those days, computers were rudimentary, and Johnson solved these intricate equations manually. Moreover, Johnson calculated the flight trajectory amidst uncertainties about how heat, gravity, atmosphere, and other factors would impact the spacecraft. Despite the remarkable complexity of these calculations, Johnson executed them with precision and grace.

Undeterred by the cultural norms of the time, Johnson challenged these norms by attending high-level meetings typically reserved for white men, paving the way for women of color in STEM. Collaborating with multiple teams, Johnson would prove to be an indispensable asset, forming bridges across diverse perspectives that pushed the collective toward innovation and advancements.

In 1962, John Glenn, one of the Mercury 7 astronauts, was getting ready for his mission to orbit the Earth. Though IBM computers had calculated the flight trajectory, Glenn was unconvinced and uttered the famous words, "Get the girl to check the numbers." He trusted the implausible intellect of Johnson over the electronic machines. Sure enough, Johnson manually verified the calculations to ensure their accuracy, successively guaranteeing Glenn's safe return to Earth.

The journey in which Katherine Johnson solved numerous complex problems and contributed to significant breakthroughs continued on. As a brilliant mathematician, she mapped the trajectory for Apollo 11 that put humanity on the moon, and she also contributed to the later Apollo 13 mission. Working deftly with her interdisciplinary team, Johnson became a prominent figure in crafting precise, robust predictions required for space flight.

Beyond her technical prowess, Johnson's journey was a lesson in tenacity, humility, and collaboration. Thriving despite adversity and discrimination, she demonstrated the power of perseverance in breaking through barriers and surpassing limitations. An often-underappreciated heroine in the space race, Johnson's accomplishments earned her a rightful place among the pantheon of innovators and explorers.

As we slip into the orbit of our present-day endeavors in research and development, the spirit of Katherine Johnson, grounded in intellect, tenacity, and collaboration, remains our compass. Unquestionably, she continues to inspire us all to reach for the stars. As we push forth into untrodden territories, may we each embody Katherine Johnson's unwavering spirit and share her unquenchable thirst for knowledge, only then can we begin to defy gravity, unhindered by the arbitrary constraints nestled within our minds.

Leading Through Uncertainty: Vision, Adaptability, and Inclusivity in High - Stakes Research Environments

Leading Through Uncertainty: Vision, Adaptability, and Inclusivity in High - Stakes Research Environments

In the high - stakes world of scientific research and discovery, uncertainty is a constant companion. As researchers push the boundaries of what is known and understood, they often venture into uncharted territory, where established rules and frameworks may no longer apply. At the forefront of this dynamic landscape are visionary leaders like Grace Hopper and Katherine Johnson, both distinguished in their respective fields for their ability to successfully navigate uncertainty, cultivating environments that foster innovation and inclusivity amidst the challenges they faced.

Grace Hopper, a pioneering computer scientist and developer of the first programming languages, embodied the spirit of fearless exploration. As a leader, she possessed the ability to envision a future where complex problems could be solved through the ingenuity of computers, and she revolutionized the field by creating the groundwork for programming languages like COBOL. Her constant commitment to pushing boundaries within her field was matched by her adaptability to the ever - evolving landscape of computer science. In an environment where new discoveries and technologies emerged at a breakneck pace, Hopper's flexibility and foresight enabled her to stay ahead of the curve and lead her team to success.

Similarly, Katherine Johnson demonstrated exceptional adaptability as a mathematician and human computer at NASA. Despite the limitations placed upon her as an African - American woman in a male - dominated field, Johnson broke through countless barriers and navigated tense political climates with grace and grit. Her ability to adapt to new challenges and fluidly adjust her mathematical models, as well as her complex calculations for orbital mechanics, proved instrumental in the success of numerous space missions during the early days of the American space program.

While both Hopper and Johnson faced unique challenges within their respective contexts, a common thread of their leadership emerges in the form of their inclusive approach. Their commitment to collaboration and a culture of inquisitiveness allowed team members from various backgrounds to flourish in high - stakes environments. Grace Hopper's groundbreaking work in

computer programming at a time when women were vastly underrepresented
in the field illustrates her willingness to challenge the status quo and foster
greater gender diversity in her field. Katherine Johnson, too, advocated
for the advancement of women in STEM and proved through her work at
NASA that inclusion and diversity lead to breakthrough successes.

At the heart of it all, leaders like Hopper and Johnson recognize that
high - stakes research environments are rich in uncertainty, but they also
demonstrate that embracing this uncertainty is key to overcoming seemingly
insurmountable odds. The combination of vision, adaptability, and inclusivity empowered them to blaze trails in the face of adversity and establish
legacies that have left indelible marks on their respective disciplines.

In the age of artificial intelligence, virtual reality, and nanotechnology,
today's research leaders must also wrestle with uncertainty as they shape the
course of humanity's progress. Emulating the spirit of Hopper and Johnson's
leadership style by incorporating a clear vision, adaptability, and inclusivity
offers a roadmap for navigating uncharted territories in the intellectual
realm.

As researchers continue to push boundaries and redefine the limits of
what is possible, they depend on leaders who are equipped to embrace
uncertainty, foster innovation, and build inclusive teams that edge closer
to solving the greatest mysteries of our time. Hopper and Johnson's legacy
challenges today's trailblazers to stand at the precipice of the unknown
and take the plunge with unwavering conviction, confident that their vision,
adaptability, and inclusive spirit will guide them as they reshape the world
in unimaginable ways.

Developing and Nurturing Talent: Lessons from Hopper and Johnson in Coaching and Mentoring

Developing and nurturing talent is essential to fostering a culture of innovation and excellence. Grace Hopper and Katherine Johnson, two exceptional
leaders in their fields of computer science and mathematics, left an indelible
impact on the world of research, not only through their pioneering work but
through their incredible ability to actualize the potential of those around
them. An examination of their successes yields valuable lessons in coaching
and mentoring, which can be applied to modern research endeavors to

optimize talent development within teams.

Quintessentially, the mentor - protégé bond is unique as it transcends
hierarchical ranks and disciplines, often cutting across lines that usually
segregate people in their respective fields. Katherine Johnson, an African
- American woman, confronted formidable racial and gender barriers as
she pursued a career in mathematics during the mid - 20th century. As
a researcher at NASA, Katherine faced isolation, marginalization, and
systemic discrimination. However, she did not allow these challenges to
stifle her passion for mentorship and knowledge sharing. Johnson formed
close relationships with her trainees, actively countering the biases by
creating an environment conducive to learning. This attribute not only
allowed her mentees to thrive, but it also advanced the efficacy of the team,
driving them towards substantial breakthroughs in space exploration.

Grace Hopper, a pioneer in computer programming, and the creator of
the first compiler, had a keen eye for talent. As a mentor, she helped shape
numerous careers by investing in and acknowledging the accomplishments
of her protégés. Hopper spent countless hours coaching her subordinates,
using her skills to explain and simplify complex ideas. As a testament to
her commitment to mentorship, she said, "You don't manage people, you
manage things. You lead people." This philosophy embodies the dynamic
essence of a mentor - protégé relationship, whereby the mentor cultivates
the talent and guides the protégé through the arduous process of research
and discovery.

Both Hopper and Johnson actively nurtured a sense of camaraderie
within their teams, inculcating an organic bond of mutual support. For
researchers embarking upon uncharted territory, driven only by curiosity and
conjecture, the journey can be isolating and frustrating. These visionary
leaders understood the need for creating a supportive environment for
collaboration, instilling faith and patience to navigate the turbulent road
of scientific inquiry. Additionally, they taught their protégés to exercise
humility and resilience in the face of setbacks.

Direct, open lines of communication are integral to mentoring relationships, paving the way for the free exchange of ideas, critique, and advice.
Hopper and Johnson both strove to be accessible to their trainees, providing
guidance even outside regular working hours. They understood the importance of being approachable to facilitate the flow of knowledge and inspire

open discussions. The strength of their mentoring hinged on their ability to communicate effectively, a skill crucial to modern mentors seeking to cultivate their protégés' talents.

Grace Hopper and Katherine Johnson exemplified the transformative power of coaching and mentoring in nurturing talent and fostering ground-breaking research. Emulating their approaches may serve as a compass for navigating the future of innovation and discovery.

In embracing their lessons, researchers and leaders can strengthen the roots of the mentor - protégé bond, allowing true brilliance to burgeon. As the landscape of human knowledge expands, so too must our capacity to share and cultivate our greatest resource - the potential that lies within every inquisitive, probing mind. And as Hopper once remarked succinctly, "The most dangerous phrase in the language is, 'We've always done it this way.'" The onus is on today's scientists, engineers, and researchers to engage in this intergenerational exchange of ideas and guidance, paving the way for newer, groundbreaking realms of exploration.

Embracing Diversity in Thought and Background: The Power of Collaboration in Fostering Innovations

Embracing diversity in thought and background may at first appear to be an inconsequential aspect of scientific research and innovation. Yet, as history has shown, it is precisely the range of perspectives and problem - solving approaches that have allowed humanity to break through barriers and make the impossible, possible.

Consider the case of the 1950s, when the world of computer science was in its nascent stages. Men and women were working side by side to build the foundations of modern computing - a rare instance of gender diversity in a world still bound by traditional roles. Among them was Grace Hopper, a female computer programmer who invented one of the first compiler tools. Hopper's unique perspective on programming languages and compilers allowed her to create a language that was more accessible and widely adopted than those of her peers.

In the same era, Katherine Johnson, a brilliant African - American mathematician, played a pivotal role in determining the trajectory of the Apollo 11 mission. Her work in calculating flight paths was critical for the mission's success and contributed significantly to the U.S. victory in the space race. Her ability to excel in a challenging field in the face of racial and gender discrimination demonstrates the power of diversity in fostering innovation.

As we move further into the 21st century, the scientific challenges we face require us to draw on ever greater resources of critical thinking, new perspectives, and fresh ideas; resources that can only be accessed by embracing diversity. We must begin by understanding that diversity is not limited to demographic representation but extends to cognitive diversity as well. Our differences and varied backgrounds can, when welcomed, inspire unique solutions.

To harness the full power of diversity, investigators must proactively break down artificial barriers to collaboration. They must learn to work effectively across disciplines, ethnicities, and cultures, forging partnerships that leverage the collective wisdom of many. This may require a willingness to question and dismantle deeply ingrained biases and assumptions. For example, leaders of research teams should consciously strive to create inclusive environments where individuals feel comfortable sharing their unique perspectives and ideas, irrespective of their cultural or educational backgrounds.

One notable example of how diversity can positively impact scientific endeavors is the Human Genome Project, an international collaboration that successfully mapped the human genome. The project, as vast as its scale and complexity, brought together scientists from various countries, disciplines, and backgrounds - working in tandem to unlock intricate genomic complexities, side - by - side.

The power of diversity in collaborative scientific efforts is also evident when examining the impact of global communication and dissemination of information throughout academia. Encyclopedic efforts like Wikipedia are a product of this digital evolution and stand as a testament to the outcomes of bringing together thousands of individuals with disparate backgrounds and expertise, working cooperatively toward a common goal.

To foster a culture of innovation, we must resist the temptation to surround ourselves with like - minded individuals who only serve to echo our thoughts and reinforce our assumptions. The "echo chamber" phenomenon, wherein people only engage with others who share similar views and opinions, can stagnate creativity and stifle groundbreaking discoveries. Instead, we must actively seek out those who think differently, who approach problems from various angles, and with whom we may not always agree.

As we conclude our examination of the power of diversity in driving innovation, it is worth noting that our efforts in embracing diversity do not end merely with assembling diverse research teams. It extends to nurturing an environment of openness, respect, and encouragement - where individuals feel valued and are inspired to contribute their unique talents and ideas effectively. Looking ahead, the onus lies on each one of us to break free from conventional silos and prejudices and create a vibrant culture of relentless inquiry and ground - breaking discovery, as the true embodiment of the legacies of visionaries such as Grace Hopper and Katherine Johnson.

Cultivating a Culture of Curiosity: Encouraging Continuous Learning and Risk - Taking

A spark of inspiration ignites in the recesses of the mind, born from a single question that initiates a cascade of ideas: "What if?" Curiosity is the driving force behind innovation, the whisper that compels us to explore the unknown, grapple with challenges, and push the boundaries of human knowledge. In the world of research, curiosity is the lifeblood that fuels groundbreaking discoveries and advancements. Atmospheres filled with curiosity cultivate creativity, nurture the progress of learning, and ultimately empower individuals and organizations to produce their most inspired work.

Creating an environment of curiosity requires more than just a fleeting fascination with knowledge; it necessitates fostering a culture where the pursuit of learning is valued and rewarded, where individuals feel empowered to explore their ideas and engage in open discussions, where taking calculated risks is encouraged rather than derided, and where setbacks and missteps pave the way for growth. The examples of Grace Hopper and Katherine Johnson, whose leadership in their respective fields were partly defined by their nurturing of curiosity, exemplify the powerful effect of curiosity in shaping the future of research.

The pursuit of continuous learning is at the heart of cultivating curiosity. A genuine thirst for knowledge and understanding, accompanied by an insatiable appetite for learning, is the seed from which creativity and innovation grow. Encouraging this passion for learning and exploration throughout all levels of an organization provides immense dividends, both in terms of individual growth and collective innovation. A culture that actively prioritizes learning, setting aside time for employees to engage with new subjects, attend conferences, or participate in workshops, demonstrates a tangible commitment to the intrinsic value of intellectual development and curiosity. By promoting the notion that there is always more to learn and discover, organizations foster a natural environment for new ideas to take root.

Equally important is harboring an open atmosphere for the exchange and elaboration of ideas - an environment that looks upon questions and alternative viewpoints as fertile ground for advancing understanding and knowledge. Facilitating this type of environment encourages individuals to think critically and actively engage with diverse perspectives, ultimately driving intellectual growth and collective problem-solving. Open discussions foster an organic sense of community and collaboration within which groundbreaking ideas can take shape.

In addition to nurturing an atmosphere of openness, it is vital to cultivate an acceptance of risk-taking. Fear of failure stifles creativity, stymieing exploration and ultimately preventing valuable opportunities for growth. Embracing risk-taking as a necessary component of innovation allows researchers to explore unconventional avenues and push beyond the boundaries of existing knowledge. As long as risks are taken from a place of calculated inquiry, even failed experiments can provide invaluable insights and hidden benefits so long as the researcher is open to lessons from the results.

The idea that setbacks can lead to growth is an essential part of creating a culture of curiosity. The journey of exploration is filled with pitfalls, and it is how we respond to these challenges that ultimately shapes our ability to forge ahead with newfound learning and understanding. High-level research environments face inherent challenges, but recognizing that setbacks are often necessary steps towards growth enables the continuous development of skills, knowledge, and methodologies.

Cultivating a culture of curiosity is not just a lofty ideal, but rather an essential part of driving innovation and advancing our understanding of the world around us. By fostering an environment where learning is valued, open discussion is encouraged, and individuals can take meaningful risks that nurture growth and progress beyond setbacks, we offer researchers the fertile ground necessary for inspired breakthroughs. If we are to follow in the footsteps of visionaries like Grace Hopper and Katherine Johnson, we must take the risk ourselves - to constantly push the boundaries of our knowledge, embrace the uncertainty of the unknown, and let curiosity guide us in our relentless pursuit of discovery. And so, in the spirit of curiosity, the question we must ask ourselves is not "What if we fail?" but rather, "What if we don't even try?"

The Legacy Continues: Applying Hopper and Johnson's Leadership Principles in Modern Research Endeavors

Great leaders leave a legacy that continues to shape future generations long after they have passed on. Grace Hopper and Katherine Johnson, two trailblazing women in the world of math, science, and computer programming, each left their own indelible marks on the fields they devoted their lives to. Just as their remarkable achievements were made through breaking down barriers and pushing the boundaries of what was thought to be possible, modern research endeavors can similarly benefit from following in their footsteps.

One of the most important principles in the leadership of Hopper and Johnson was their steadfast commitment to inclusivity in the research environment. As both women battled discrimination on various grounds throughout their careers, they understood the value of bringing together diverse perspectives and encouraging collaboration. Today's researchers can actively support inclusivity in their teams by fostering an environment that values and respects every individual's contributions, recognizing that the most innovative solutions emerge when diverse minds come together to tackle complex challenges.

Another core tenet of Hopper's and Johnson's leadership was their unwavering focus on curiosity. They were passionate and driven by an insatiable thirst for discovery, and this relentless pursuit of knowledge was an essential aspect of their successes. Embracing a culture of curiosity within modern research endeavors can empower teams to continuously question, learn, and adapt, cultivating an atmosphere that stimulates creative problem - solving and breakthrough innovations. Encouraging questions and open dialogue, as well as supporting professional development opportunities for team members, ensures that this spirit of curiosity is sustained and nurtured.

A remarkable aspect of Hopper's and Johnson's careers was their ability to forge a path through uncertainty and lead by example in times of transformative change. Both women made significant strides despite having limited guidance or precedent in their respective fields, which, in turn, inspires modern-day researchers to embrace the unknown with courage and aplomb. By cultivating an atmosphere where risk-taking is supported and encouraged, research teams can navigate uncharted territories and discover unprecedented solutions to pressing challenges.

Furthermore, both Hopper and Johnson understood the importance of adapting to ever-evolving technological landscapes and the necessity of staying at the forefront of their fields. As technology continues to advance at a rapid pace, this adaptability becomes even more crucial in modern research. Leaders who foster an environment where continuous learning and skills development are prioritized position their teams at the vanguard of their industries, armed with the cutting-edge knowledge necessary to bring about transformative breakthroughs.

Lastly, Hopper and Johnson were adamant about nurturing talent and developing the potential of those around them. As trailblazers in their respective disciplines, each of these pioneering women understood the importance of mentorship in fostering the next generation of leaders. By investing in the personal and professional growth of their teams, leaders in modern research initiatives can ensure that the spark of innovation and discovery remains alight for future generations to come.

In the constant pursuit of progress, the remarkable legacies of Grace Hopper and Katherine Johnson offer invaluable leadership principles that continue to fuel groundbreaking research initiatives. As we embark on the next frontier of discovery, it is essential that we take up their torch and embody the spirit with which they approached their extraordinary feats: with unrelenting curiosity, boundless adaptability, and fierce dedication to the untapped potential of collaboration and inclusivity.

In igniting this spirit, as Hopper and Johnson did, we can forge forth into uncharted territories of innovation - not only breaking through barriers but creating new paradigms for what is deemed possible in the vast expanse that lies just beyond the horizon.

www.ingramcontent.com/pod-product-compliance
Lightning Source LLC
Chambersburg PA
CBHW062315290526
45794CB00005B/1812